Undergraduate Texts in Mathematics

Undergraduate Texts in Mathematics

Apostol: Introduction to Analytic Number Theory.
1976. xii, 338 pages. 24 illus.

Armstrong: Basic Topology.
1983. xii, 260 pages. 132 illus.

Bak/Newman: Complex Analysis.
1982. x, 224 pages. 69 illus.

Banchoff/Wermer: Linear Algebra Through Geometry.
1983. x, 257 pages. 81 illus.

Childs: A Concrete Introduction to Higher Algebra.
1979. xiv, 338 pages. 8 illus.

Chung: Elementary Probability Theory with Stochastic Processes.
1975. xvi, 325 pages. 36 illus.

Croom: Basic Concepts of Algebraic Topology.
1978. x, 177 pages. 46 illus.

Curtis: Linear Algebra: An Introductory Approach (4th edition)
1984. x, 337 pages. 37 illus.

Dixmier: General Topology.
1984. x, 140 pages. 13 illus.

Ebbinghaus/Flum/Thomas Mathematical Logic.
1984. xii, 216 pages. 1 illus.

Fischer: Intermediate Real Analysis.
1983. xiv, 770 pages. 100 illus.

Fleming: Functions of Several Variables. Second edition.
1977. xi, 411 pages. 96 illus.

Foulds: Optimization Techniques: An Introduction.
1981. xii, 502 pages. 72 illus.

Foulds: Combinatorial Optimization for Undergraduates.
1984. xii, 222 pages. 56 illus.

Franklin: Methods of Mathematical Economics. Linear and Nonlinear Programming. Fixed-Point Theorems.
1980. x, 297 pages. 38 illus.

Halmos: Finite-Dimensional Vector Spaces. Second edition.
1974. viii, 200 pages.

Halmos: Naive Set Theory.
1974, vii, 104 pages.

Iooss/Joseph: Elementary Stability and Bifurcation Theory.
1980. xv, 286 pages. 47 illus.

Jänich: Topology
1984. ix, 180 pages (approx.). 180 illus.

Kemeny/Snell: Finite Markov Chains.
1976. ix, 224 pages. 11 illus.

Lang: Undergraduate Analysis
1983. xiii, 545 pages. 52 illus.

Lax/Burstein/Lax: Calculus with Applications and Computing, Volume 1. Corrected Second Printing.
1984. xi, 513 pages. 170 illus.

LeCuyer: College Mathematics with A Programming Language.
1978. xii, 420 pages. 144 illus.

Macki/Strauss: Introduction to Optimal Control Theory.
1981. xiii, 168 pages. 68 illus.

continued after Index

L. R. Foulds

Combinatorial Optimization for Undergraduates

With 56 Illustrations

Springer-Verlag
New York Berlin Heidelberg Tokyo

L. R. Foulds
Department of Industrial and Systems Engineering
University of Florida
Gainesville, FL 32611
U.S.A.

AMS Classifications: 05-01, 49-01

Library of Congress Cataloging in Publication Data
Foulds, L. R.
Combinatorial optimization for undergraduates.
(Undergraduate texts in mathematics)
Bibliography: p.
Includes index.
1. Combinatorial optimization. I. Title.
II. Series.
QA164.F68 1984 519 84-5381

Typeset by Science Typographers, Inc., Medford, New York.
Printed and bound by R. R. Donnelley & Sons, Harrisonburg, Virginia.
Printed in the United States of America.

9 8 7 6 5 4 3 2 1

ISBN 0-387-90977-X Springer-Verlag New York Berlin Heidelberg Tokyo
ISBN 3-540-90977-X Springer-Verlag Berlin Heidelberg New York Tokyo

To my parents, Edith and Reginald,
of Ruakaka, New Zealand

Preface

The major purpose of this book is to introduce the main concepts of discrete optimization problems which have a finite number of feasible solutions. Following common practice, we term this topic *combinatorial optimization*. There are now a number of excellent graduate-level textbooks on combinatorial optimization. However, there does not seem to exist an undergraduate text in this area. This book is designed to fill this need.

The book is intended for undergraduates in mathematics, engineering, business, or the physical or social sciences. It may also be useful as a reference text for practising engineers and scientists. The writing of this book was inspired through the experience of the author in teaching the material to undergraduate students in operations research, engineering, business, and mathematics at the University of Canterbury, New Zealand. This experience has confirmed the suspicion that it is often wise to adopt the following approach when teaching material of the nature contained in this book. When introducing a new topic, begin with a numerical problem which the students can readily understand; develop a solution technique by using it on this problem; then go on to general problems. This philosophy has been adopted throughout the book. The emphasis is on plausibility and clarity rather than rigor, although rigorous arguments have been used when they contribute to the understanding of the mechanics of an algorithm. An example of this is furnished by the construction of the labeling method for the maximal-network-flow problem from the proof of the max-flow, min-cut theorem.

The book comprises two parts—Part I: Techniques and Part II: Applications. Part I begins with a motivational chapter which includes a description of the general combinatorial optimization problem, important current problems, a description of the fundamental algorithm, a discussion of the

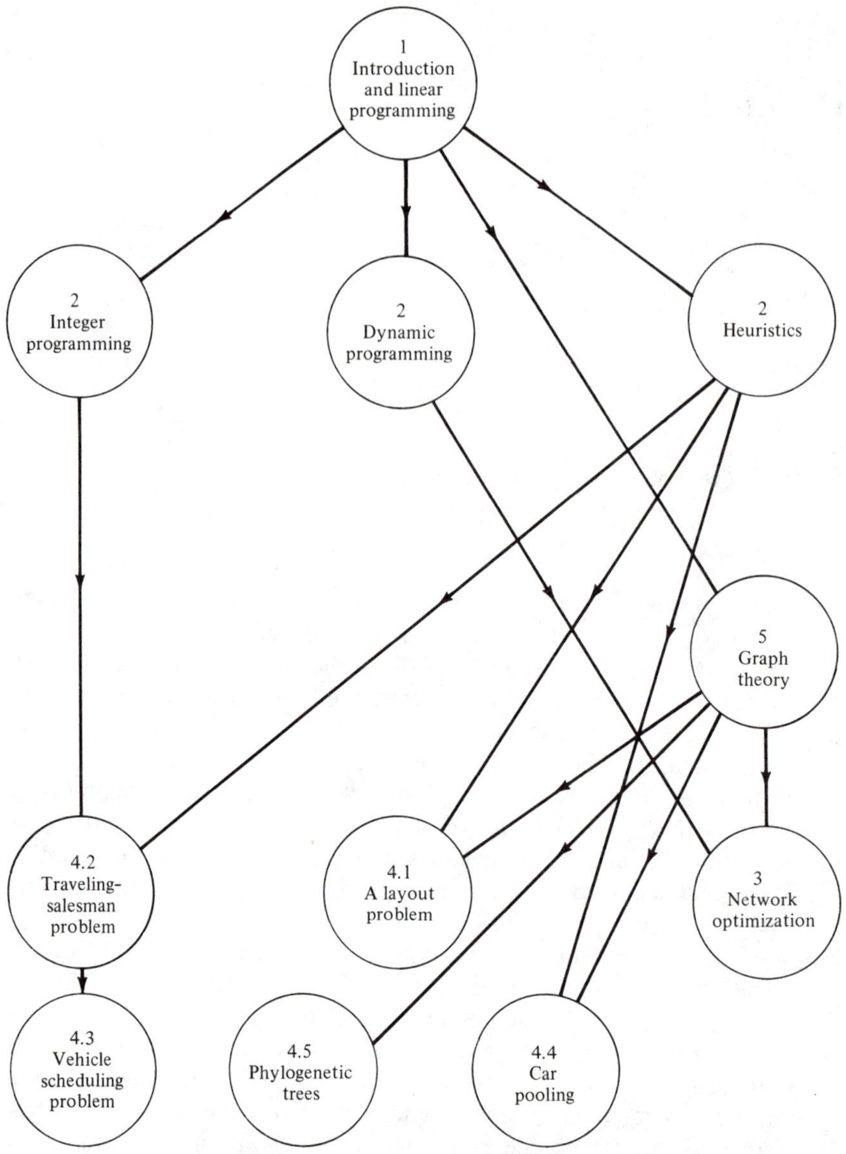

Plan of the Book.

need for efficient algorithms, and the effect of the advent of the digital computer. This is followed by a chapter on linear programming and its extensions. Chapter 2 describes the basic procedures of three of the most important combinatorial optimization techniques—integer programming, dynamic programming, and heuristic methods. Chapter 3 is concerned with optimization on graphs and networks.

Part II poses a variety of problems from many different disciplines—the traveling-salesman problem, the vehicle scheduling problem, car pooling, evolutionary tree construction, and the facilities layout problem. Each problem is analyzed and solution procedures are then presented. Some of these procedures have never appeared before in book form.

The book contains a number of exercises which the reader is strongly encouraged to try. Mathematics is not a spectator sport! These exercises range from routine numerical drill-type exercises to open questions from the research literature. The more challenging problems have an asterisk preceding them. The author is grateful for this opportunity to express his thanks for the support he received from the University of Canterbury while writing this book, and to his doctoral student John Giffin, who contributed to Section 4.1. He is also extremely thankful to his wife Maureen, who not only provided enthusiastic encouragement, but also typed the complete manuscript. Finally, the author pays a hearty tribute to the staff at Springer-Verlag New York for their patience, skill, and cooperation during the preparation of this book.

Gainesville, Florida L. R. FOULDS

Contents

PART I

TECHNIQUES

CHAPTER 0

Introduction to the Techniques of Combinatorial Optimization

0.1. The General Problem

Optimization is concerned with finding the best (or optimal) solution to a problem. In this book we are concerned with problems that can be stated in an unambiguous way, usually in terms of mathematical notation and terminology. It is also assumed that the value of any solution to a given problem can be measured in a quantifiable way and this value can be compared with that of any other solution to the problem. Problems of this nature have been posed since the beginning of mankind. One of the earliest is recorded by Virgil in his *Aeneid* where he relates the dilemma of Queen Dido, who was to be given all the land she could enclose in the hide of a bull. She cut the hide into thin strips and joining these together formed a semicircle within which she enclosed a sizeable portion of land with the Mediterranean coast as the diameter. Later Archimedes conjectured that her mathematical solution was optimal; that is, a semicircle is the curve of fixed length which will, together with a straight line, enclose the largest possible area. This conjecture can be proved using a branch of optimization called the calculus of variations.

The problem just described has an infinite number of solutions as there is an infinite number of possible curves of any given length. However, there is an important class of optimization problems which have only a finite number of solutions. The body of knowledge concerned with the theory and techniques for these problems is called "combinatorial optimization" and it is with this class that our book deals. Let S be the finite set of solutions to a problem and assume each solution $x \in S$, can be evaluated and assigned a real number $f(x)$ indicating its worth. This assignment may be in terms of some benefit, such as profit, which is to be maximized, or some detriment, such as cost, which is to be minimized.

We now formally introduce the general problem of combinatorial optimization. Let

$$f: D \to \mathbb{R}$$

be a real-valued function with domain D. Let

$$S \subseteq D.$$

Definition 0.1. $x^* \in S$ is a *global maximum of* f if

$$f(x^*) \geq f(x) \qquad \text{for all } x \in S.$$

The definition of a global minimum is analogous.

Definition 0.2. $x^* \in S$ is a *global extremum of* f if x^* is either a global maximum or global minimum of f.

The "general maximization problem of combinatorial optimization" is to find x^* such that x^* is a global maximum of f; that is, to identify x^* such that

$$f(x^*) = \text{Max}(f(x)) \qquad x \in S.$$

The definition of the "general minimization problem of combinatorial optimization" is analogous.

S is called the set of *feasible* solutions and if

$$x \in S,$$

x is called a *feasible* solution or is termed *feasible*. If $\bar{x} \in D$ and

$$f(\bar{x}) \geq f(x) \qquad \text{for all } x \in S,$$

$\bar{x}$ is termed an *upper bound for f on S*. If

$$f(\bar{x}) \leq f(x) \qquad \text{for all } x \in S,$$

$\bar{x}$ is termed a *lower bound for f on S*. If $\bar{x}$ is an upper bound for f on S and

$$f(\bar{x}) \leq f(x)$$

for all upper bounds x for f on S, then $\bar{x}$ is termed a *least upper bound for f on S*. If $\bar{x}$ is a lower bound for f on S and

$$f(\bar{x}) \geq f(x)$$

for all lower bounds x for f on S, then $\bar{x}$ is termed a *greatest lower bound for f on S*.

Note that $\bar{x}$ may or may not be a member of S. Of course if

$$\bar{x} \in S,$$

$\bar{x}$ is a global extremum of f.

0.1.1. An Illustrative Example of a Combinatorial Optimization Problem: The Shortest Hamiltonian Path

This section is based upon an article by D. F. Robinson appearing in the Proceedings of the First Australian Conference on Combinatorics, Newcastle, Australia, August 1972.

Of course the nature of S can vary considerably from one problem to another. Even though S is finite it may be extremely large and further it may not be an easy task to identify its elements. We now present an illustrative example of a combinatorial optimization problem that is simple in concept, in order to give the reader some idea of what is to come.

Let $V=\{v_1, v_2, \ldots, v_n\}$ be a set of n cities where $n>1$. Consider the problem of finding a shortest itinerary which passes through all the cities of V. Let d_{ij} represent the distance from v_i to v_j, $1 \le i \le n$, $1 \le j \le n$. The distance matrix

$$\mathbf{D} = \left[d_{ij}\right]_{n \times n}$$

is assumed to be symmetric in the sense that

$$d_{ij} = d_{ji} \qquad 1 \le i \le n, 1 \le j \le n.$$

This problem is similar to one in the literature known as the traveling-salesman problem which is the subject of Section 4.2.

As V has n members there are $n!$ paths. We express a typical path as

$$x = \langle v_{\alpha(1)}, v_{\alpha(2)}, \ldots, v_{\alpha(n)} \rangle,$$

where $\{v_{\alpha(1)}, v_{\alpha(2)}, \ldots, v_{\alpha(n)}\} = \{v_1, v_2, \ldots, v_n\}$, and x is the path which begins at $v_{\alpha(1)}$ and then visits $v_{\alpha(2)}$, $v_{\alpha(3)}$, and so on, ending at $v_{\alpha(n)}$. The set S of solutions to this problem is

$$S = \{\langle v_{\alpha(1)}, v_{\alpha(2)}, \ldots, v_{\alpha(n)} \rangle : \{\alpha(1), \alpha(2), \ldots, \alpha(n)\} = \{1, 2, \ldots, n\}\}.$$

If $x = \langle v_{\alpha(1)}, v_{\alpha(2)}, \ldots, v_{\alpha(n)} \rangle \in S$, then the value of x, $f(x)$ is the length of x,

$$f(x) = \sum_{i=1}^{n-1} d_{\alpha(i), \alpha(i+1)}.$$

Then the problem is find

$$f(x^*) = \min_{x \in S} \{ f(x) \}.$$

To each path $x = \langle v_{\alpha(1)}, v_{\alpha(2)}, \ldots, v_{\alpha(n)} \rangle$ there corresponds a reverse path, $x^R = \langle v_{\alpha(n)}, v_{\alpha(n-1)}, \ldots, v_{\alpha(1)} \rangle$. Because d is symmetric, $f(x) = f(x^R)$ for all $x \in S$. Hence the minimum path will not be unique.

We define the following elementary operations on a path:

(i) Break the path $\langle v_{\alpha(1)}, v_{\alpha(2)}, \ldots, v_{\alpha(n)} \rangle$ after some point $v_{\alpha(m-1)}$ $(2 \le m \le n)$ and join $v_{\alpha(n)}$ to $v_{\alpha(1)}$. The new path is

$$\langle v_{\alpha(m)}, v_{\alpha(m+1)}, \ldots, v_{\alpha(n)}, v_{\alpha(1)}, v_{\alpha(2)}, \ldots, v_{\alpha(m-1)} \rangle.$$

This new path will be shorter than the original one if and only if

$$d_{\alpha(n), \alpha(1)} < d_{\alpha(m-1), \alpha(m)}.$$

(ii) Break the path $\langle v_{\alpha(1)}, v_{\alpha(2)}, \ldots, v_{\alpha(n)} \rangle$ after some point $v_{\alpha(m-1)}$ $(2 \le m \le n-1)$ and then reverse the direction of the second part to yield

$$\langle v_{\alpha(1)}, v_{\alpha(2)}, \ldots, v_{\alpha(m-1)}, v_{\alpha(n)}, v_{\alpha(n-1)}, \ldots, v_{\alpha(m)} \rangle.$$

This new path will be shorter than the original one if and only if

$$d_{\alpha(m-1), \alpha(n)} < d_{\alpha(m-1), \alpha(m)}.$$

(iii) The reverse of operation (ii). Break the path $\langle v_{\alpha(1)}, v_{\alpha(2)}, \ldots, v_{\alpha(n)} \rangle$ after $v_{\alpha(m)}$ $(2 \le m \le n-1)$ and reverse the first half to give the path

$$\langle v_{\alpha(m)}, v_{\alpha(m-1)}, \ldots, v_{\alpha(1)}, v_{\alpha(m+1)}, v_{\alpha(m+2)}, \ldots, v_{\alpha(n)} \rangle.$$

This new path is shorter than the original if

$$d_{\alpha(1), \alpha(m+1)} < d_{\alpha(m), \alpha(m+1)}.$$

(iv) Take any pair of adjacent points $v_{\alpha(m)}, v_{\alpha(m+1)}$ in the path

$$\langle v_{\alpha(1)}, v_{\alpha(2)}, \ldots, v_{\alpha(n)} \rangle$$

and reverse their order to obtain

$$\langle v_{\alpha(1)}, v_{\alpha(2)}, \ldots, v_{\alpha(m-1)}, v_{\alpha(m+1)}, v_{\alpha(m)}, v_{\alpha(m+2)}, \ldots, v_{\alpha(n)} \rangle.$$

The cases $m = 1$ or $m = n-1$ have been dealt with in (ii) and (iii). Otherwise the new path is shorter than the original if

$$d_{\alpha(m-1), \alpha(m+1)} + d_{\alpha(m), \alpha(m+2)} < d_{\alpha(m+2), \alpha(m-1)} + d_{\alpha(m), \alpha(m+1)}.$$

We now note some properties of these operations:

(a) If it is possible to obtain path y by an elementary operation on a path x, then x can be obtained from y by an elementary operation.
(b) If a path y can be obtained from a path x by an elementary operation, then path y^R, the reverse of y, can be obtained from x^R by an elementary operation.
(c) Each path can be considered a permutation on $\{v_1, v_2, \ldots, v_n\}$. Type (iv) operations are in effect permutation transpositions.

It can be shown that any permutation can be expressed as a product of transpositions. Hence any path can be transformed into any other path by a finite sequence of type (iv) operations. Hence any pair of paths can be transformed, one into the other, by a finite sequence of operations of types (i)–(iv).

Table 0.1. A City-to-City Distance Table

	v_1	v_2	v_3	v_4	v_5	v_6	v_7	v_8	v_9	v_{10}
v_2	201									
v_3	428	227								
v_4	207	156	348							
v_5	232	159	351	25						
v_6	564	363	136	448	423					
v_7	73	274	501	186	211	634				
v_8	19	220	447	226	251	583	72			
v_9	508	307	176	340	314	118	526	527		
v_{10}	302	101	126	222	225	262	375	321	217	
v_{11}	165	210	437	65	90	513	144	184	405	311

Let us now search for a path of minimum length. The ideas will be made more clear by examining a numerical example. Table 0.1 gives the distances between a set of 11 cities.

In Table 0.2 we set out the successive shorter paths, represented by the cities in order and the distances between them. The starting path A_0 consists of the cities in order of increasing subscript. The greatest distance between successive cities is 634 (from v_6 to v_7). This is greater than $d_{1,11}$ $(=165)$. We therefore split the path A_0 between v_6 and v_7 and join v_1 to v_{11} by a type (i) operation. We denote this break by the symbol $\wedge$ in the appropriate place. The new path is denoted by A_1. We can never use type (i) operations twice successively to any advantage. We now turn to type (ii) operations and compare the distances between cities with the distances from the right-hand side of a pair to v_7. We find that v_7 is closer to v_{11} than v_{10} is. So we form path A_2 by reversing the section of the path from v_7 to v_{10}. This is denoted by the symbols $<$ and $>$. In A_2 the cities v_9 and v_8 are 527 apart. This exceeds the distance between v_{10} and v_6. So a type (i) operation will reduce this length. We continue using type (i) and (ii) operations until we reach A_{13}. No further type (i) or (ii) operation will reduce the length of this path. A type (iii) operation, reversing the order of the last three cities, will reduce the length of A_{13}. This produces path A_{14}. No operation of any type will reduce the length of A_{14}, which is 965.

Note that the distance from v_6 to v_7 is 634. Hence it is evident that any shorter path must have one end "close" to v_6 and the other "close" to v_7. It is then a simple matter to prove that A_{14} (with its reverse) is a global minimum.

Definition 0.3. If, for all such x_i

$$f(x_0) \leq f(x_i),$$

x_0 is said to be a *local minimum of* f.

Table 0.2. A Succession of Shorter Paths

A_0	v_1	201	v_2	227	v_3	348	v_4	25	v_5	423	v_6	634	v_7	72	v_8	527	v_9	217	v_{10}	311	v_{11}
												$\wedge$									
A_1	v_7	72	v_8	527	v_9	217	v_{10}	311	v_{11}	165	v_1	201	v_2	227	v_3	348	v_4	25	v_5	423	v_6
	$<$							$>$													
A_2	v_{10}	217	v_9	527	v_8	72	v_7	144	v_{11}	165	v_1	201	v_2	227	v_3	348	v_4	25	v_5	423	v_6
				$\wedge$																	
A_3	v_8	72	v_7	144	v_{11}	165	v_1	201	v_2	227	v_3	348	v_4	25	v_5	423	v_6	262	v_{10}	217	v_9
	$<$					$>$															
A_4	v_{10}	144	v_7	72	v_8	19	v_1	201	v_2	227	v_3	348	v_4	25	v_5	423	v_6	262	v_{10}	217	v_9
																$\wedge$					
A_5	v_6	262	v_{10}	217	v_9	405	v_{11}	144	v_7	72	v_8	19	v_1	201	v_2	227	v_3	348	v_4	25	v_5
	$<$			$>$																	
A_6	v_{10}	262	v_6	118	v_9	405	v_{11}	144	v_7	72	v_8	19	v_1	201	v_2	227	v_3	348	v_4	25	v_5
						$\wedge$															
A_7	v_{11}	144	v_7	72	v_8	19	v_1	201	v_2	227	v_3	348	v_4	25	v_5	225	v_{10}	262	v_6	118	v_9
	$<$											$>$									
A_8	v_3	227	v_2	201	v_1	19	v_8	72	v_7	144	v_{11}	65	v_4	25	v_5	225	v_{10}	262	v_6	118	v_9
																		$\wedge$			
A_9	v_6	118	v_9	176	v_3	227	v_2	201	v_1	19	v_8	72	v_7	144	v_{11}	65	v_4	25	v_5	225	v_{10}
	$<$			$>$																	
A_{10}	v_9	118	v_6	136	v_3	227	v_2	201	v_1	19	v_8	72	v_7	144	v_{11}	65	v_4	25	v_5	225	v_{10}
						$\wedge$															
A_{11}	v_2	201	v_1	19	v_8	72	v_7	144	v_{11}	65	v_4	25	v_5	225	v_{10}	217	v_9	118	v_6	136	v_3
	$<$													$>$							
A_{12}	v_5	25	v_4	65	v_{11}	144	v_7	72	v_8	19	v_1	201	v_2	101	v_{10}	217	v_9	118	v_6	136	v_3
	$<$											$>$									
A_{13}	v_1	19	v_8	72	v_7	144	v_{11}	65	v_4	25	v_5	159	v_2	101	v_{10}	217	v_9	118	v_6	136	v_3
																$<$					$>$
A_{14}	v_1	19	v_8	72	v_7	144	v_{11}	65	v_4	25	v_5	159	v_2	101	v_{10}	126	v_3	136	v_6	118	v_9

Definition 0.4. If, for all such x_i

$$f(x_0) \geq f(x_i),$$

x_0 is said to be a *local maximum of* f.

Definition 0.5. If x_0 is either a local minimum of f or a local maximum of f then x_0 is said to be a *local extremum of* f.

We may generalize the above approach. Consider the general minimization problem of combinatorial optimization as defined in Section 0.1. We can define on S a collection of elementary operations with the following properties:

(a) If $y \in S$ can be obtained from $x \in S$ by an elementary operation, then x can be obtained from y by an elementary operation.
(b) Given any two $x, y \in S$ there is a finite sequence of elementary operations which convert x into y.

The elementary operations thus define a connected graph, G (see Section 5.2), whose vertices are the members of S and whose edges join members of S linked by an elementary operation. A solution process can be constructed as follows. Begin at an arbitrary vertex $x_0 \in S$ and evaluate $f(x_0)$. We then evaluate $f(x_i)$ for each x_i adjacent to x_0 in G.

If no such local minimum is detected, choose an $x_j \in S$ adjacent to x_0 for which

$$f(x_0) > f(x_j) \tag{0.1}$$

and repeat the above process with x_0 replaced by x_j. One method of selecting x_j at each stage is to choose the first member of S adjacent to x_0 for which it is discovered that (0.1) holds.

The above process must terminate in the identification of a local minimum in a finite number of steps as S is finite. A possible minor complication may arise in that a "plateau" x_p may be detected where

$$f(x_p) = f(x_j)$$

for some x_j adjacent to x_p, but with no adjacent vertex x_k such that

$$f(x_p) > f(x_k).$$

If this situation arises, the set S' of all such vertices x_j is progressively examined in the hope that a vertex x_s may be found which is adjacent to a vertex $x_j \in S'$ and

$$f(x_p) = f(x_j) > f(x_s).$$

Then the process is repeated with x_0 replaced by x_s.

Some exercise of judgment is needed in the application of this process. If the number of vertices adjacent to any given vertex is usually relatively small, local but not global minima will frequently be found. On the other hand, if the number is usually relatively large, the search procedure may take an inordinate amount of time.

0.2. Important Combinatorial Optimization Problems

This section comprises a description of a number of important combinatorial optimization problems. The list is by no means exhaustive, but is meant to give a flavor of the area. Analyses of these and other problems and references to further work done on them will be found later in the book.

0.2.1. The Minimal Cost Network Problem

Given a network with arc costs and capacities, what is the minimum cost flow assignment which transports a given commodity from source to sink?

0.2.2. The Transportation Problem

Given a distribution system from a set of warehouses to a set of factories, what is the least transportation cost assignment of a single commodity satisfying factory production capacity and warehouse demand?

0.2.3. The Facilities Layout Problem

Given a set of facilities to be laid out on a plane factory floor, what spatial arrangement of these facilities maximizes the benefit of pairwise adjacency?

0.2.4. The Traveling-Salesman Problem

Given a set of cities, what circuit of them should a salesman tour in order to minimize total distance traveled if he is to visit each city in the set, returning to his starting point?

0.2.5. The Vehicle Scheduling Problem

Given a set of vehicles to be used for servicing a number of locations, what set of tours should be assigned to the vehicles which minimizes distance traveled and services the locations subject to vehicle capacity?

0.3. The Fundamental Algorithm, Efficiency and the Digital Computer

Consider the general maximization problem of combinatorial optimization, as defined in Section 0.1. How could a specific instance of this problem be solved? That is, suppose we are given the definition of a particular finite set S and a function f such that

$$f: S \to \mathbb{R}.$$

Of course f may be nothing more than the specification of some experiment or task that produces a unique real number $f(x)$ for each $x \in S$. Then our problem is to find an $x^* \in S$ such that $f(x^*)$ is no less than $f(x)$ for any other $x \in S$. Since S is finite, we could simply evaluate $f(x)$ for every single $x \in S$ and compare all the function values and choose the largest. The element (or elements) in S corresponding to this largest value provide us with the desired solution. This approach is called the "fundamental algorithm of combinatorial optimization" and is stated as follows:

> Evaluate $f(x)$, for all $x \in S$. Choose x^* to be the value of x for which $f(x^*) \geq f(x)$, for all $x \in S$.

Naturally, there is a minimization version of this algorithm in which one chooses the smallest function value. Although, this algorithm will *theoretically* solve any finite-optimization problem, for many realistic problems the set S may have too many elements for this method to be of any practical use.

As an example of this rather depressing fact consider the Hamiltonian path problem of Section 0.1.1. Suppose the number of cities $n = 21$. Then we could consider evaluating all 21! possible itineraries and choosing the shortest. However, even if we had a computer capable of evaluating the length of one path every nanosecond it would still take over 16,000 years of continuous calculation!

This example amply demonstrates the need to search for techniques that are more efficient than simple exhaustive enumeration. Since the beginning of mathematics there has been a steady stream of such techniques proposed. Many of the earlier ideas, although far better than complete enumeration, still involved far too much calculation to be useful to a person armed only with a pencil and paper. However, the advent of the modern digital computer, with its awesome capacity to perform arithmetical calculations at astonishing speed, has rekindled interest in many of these long-forgotten techniques. This book documents some of the broad approaches that the computer has now made meaningful.

CHAPTER 1

Linear Programming and Extensions

1.1. An Introduction to Linear Programming

One of the areas of mathematics which has extensive use in combinatorial optimization is called *linear programming* (LP). It derives its name from the fact that the LP problem is an optimization problem in which the objective function and all the constraints are linear. Many real-world problems can be formulated in this way. Even more problems can be effectively approximated by an LP model. Also an LP solution method can be used as a subroutine in solving integer-programming problems (as indicated in Section 2.1) and certain nonlinear optimization problems.

In this chapter we introduce the basic LP concepts. An efficient LP algorithm, called the "simplex method," will be detailed. We also discuss the idea of a dual LP problem with a view to developing the dual simplex method. This latter method is useful in the solution of integer-programming problems as shown in Chapter 2. We begin with a simple numerical example of an LP problem.

1.1.1. A Simple LP Problem

The Melt in Your Mouth ice-cream shop makes two kinds of ice cream—nut cassata (C) and fruit pistachio (P). The shop is located in a busy tourist area and is in the fortunate position of being able to sell all the ice cream it can make. One cone of C sells for $0.75 and one cone of P sells for $0.60. A cone of C requires 4 g of fruit mix and 2 g of nuts. A cone of P requires 6 g of fruit mix and 1 g of crushed nuts. However, only 96 g of fruit mix and 24 g of crushed nuts can be produced hourly. How many of each type of cone should be made in order to maximize hourly revenue?

In order to answer this question, we first formulate the problem mathematically. We begin by defining some decision variables. Let x_1 represent the number of cones of C produced hourly and x_2 the number of cones of P produced hourly. We also define a dependent variable. Let Z be the revenue gained from producing x_1 cones of C and x_2 cones of P.

If x_1 cones of C are produced hourly and the revenue is \$0.75 per cone, then the hourly revenue for C is

$$\$0.75x_1.$$

Similarly, if x_2 cones of P are produced hourly and the revenue is \$0.60 per cone, then the hourly revenue for P is

$$\$0.60x_2.$$

So for an hourly production schedule of x_1 and x_2 cones of C and P, respectively, the total hourly revenue (in dollars) is

$$Z = 0.75x_1 + 0.60x_2.$$

We wish to find values for x_1 and x_2 which maximize this expression. Naturally, there are constraints on the values that x_1 and x_2 can feasibly assume. For instance, consider the restriction on available fruit. If x_1 cones of C are produced and each cone requires 4 g of fruit, then the total fruit required for C is

$$4x_1 \text{ g}.$$

Similarly, if x_2 cones of P are produced and each cone requires 6 g of fruit, then the total fruit required for P is $6x_2$ g. Hence the total fruit required for an (x_1, x_2) schedule is

$$4x_1 + 6x_2.$$

But there are only 96 g of fruit available. Thus

$$4x_1 + 6x_2 \leq 96.$$

We can formulate a similar constraint for the restriction on the availability of nuts. This appears below.

The complete problem is

$$\text{Maximize } 0.75x_1 + 0.60x_2 \quad (= Z) \tag{1.1}$$

subject to

$$4x_1 + 6x_2 \leq 96, \tag{1.2}$$

$$2x_1 + x_2 \leq 24, \tag{1.3}$$

$$x_1 \text{ and } x_2 \geq 0. \tag{1.4}$$

These expressions can be explained as follows:

(1.1) The objective is to maximize the hourly revenue Z.
(1.2) A maximum of 96 g of fruit mix is available hourly.
(1.3) A maximum of 24 g of nuts is available hourly.
(1.4) A nonnegative number of cones of C and P must be produced.

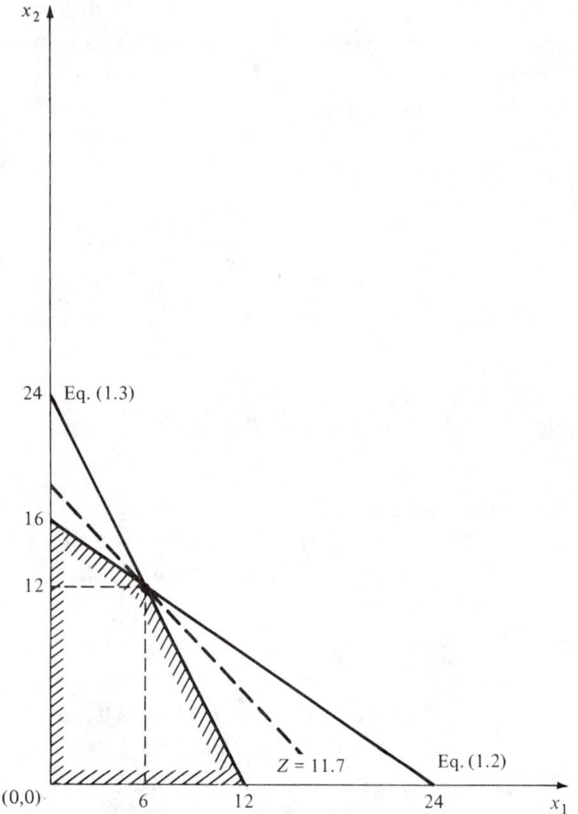

Figure 1.1. The Graphical Solution to the Example Problem.

We can solve this problem with the aid of a diagram because there are only two decision variables. The diagram is shown in Fig. 1.1. The inequalities (1.2) and (1.3) are drawn as shown along with the x_1 and x_2 axes, which represent the nonnegativity conditions $x_2 \geq 0$ and $x_1 \geq 0$, respectively. The shaded region contains all points which correspond to feasible solutions. In order to find the point representing the optimal solution, we set Z equal to an arbitrary value, for example, 10. This creates the equation

$$0.75x_1 + 0.60x_2 = 10.$$

As Z is increased this line moves parallel to itself away from the origin. When $Z = 11.7$ it coincides with a single point of the feasible region. Clearly, this point represents the optimal solution: $(x_1^*, x_2^*) = (6, 12)$. The best hourly revenue that the shop can hope for is \$11.70 from a policy of making 6 cones of C and 12 cones of P.

Suppose the amounts of fruit mix and nuts available per hour are 101 g and 25.5 g, respectively. The new optimal solution is $(x_1^*, x_2^*) = (6.5, 12.5)$. (Check this.) Obviously, it would be ludicrous to advise the shop to make an extra half ice cream cone per hour. In this case, the suggested policy would be to make 13 and 25 cones of C and P, respectively, every 2 hours. By taking lowest common multiples of optimal values in the way illustrated, feasible integer solutions can sometimes be deduced if necessary. When the lowest common multiples are too large for the corresponding solution to be practical, special techniques called *integer programming methods* must be used to solve the problem. Some simple integer programming approaches are explained in Section 2.1.

Most realistic problems have many more than two decision variables. It is difficult to draw a suitable diagram for a three-variable problem and impossible in higher dimensions. We need a more useful method for solving LP problems. Before introducing such a method let us define the general LP problem.

1.1.2. The General LP Problem

The numerical problem discussed in the last section is an example of an LP problem. In its general form, the LP problem is

$$\text{Maximize } c_1x_1 + c_2x_2 + \cdots + c_nx_n \quad (=Z)$$

subject to

$$\begin{aligned}
a_{11}x_1 + \ a_{12}x_2 + \cdots + a_{1n}x_n &\le b_1,\\
a_{21}x_1 + \ a_{22}x_2 + \cdots + a_{2n}x_n &\le b_2,\\
&\vdots\\
a_{m1}x_1 + a_{m2}x_2 + \cdots + a_{mn}x_n &\le b_m,\\
x_1, x_2, \ldots, x_n &\ge 0.
\end{aligned}$$

In the ice-cream-shop example of Section 1.1.1,

$$\begin{aligned}
&\text{the number of variables } n = 2,\\
&\text{the number of constraints } m = 2,\\
&(c_1, c_2) = (0.75, 0.60),\\
&\begin{pmatrix} b_1 \\ b_2 \end{pmatrix} = \begin{pmatrix} 96 \\ 24 \end{pmatrix},\\
&\begin{pmatrix} a_{11} & a_{12} \\ a_{21} & a_{22} \end{pmatrix} = \begin{pmatrix} 4 & 6 \\ 2 & 1 \end{pmatrix}.
\end{aligned}$$

From the way in which we have expressed these data it is clear that we can formulate the LP problem in terms of matrices as

$$\text{Maximize } \mathbf{CX} \qquad (=Z)$$

subject to

$$\begin{aligned}
\mathbf{AX} &\le \mathbf{b} \quad \text{(the constraint sets)},\\
\mathbf{X} &\ge \mathbf{0} \quad \text{(the nonnegativity conditions)}.
\end{aligned}$$

(See Section 5.1 for an explanation of the linear algebra needed to follow the rest of this chapter.) Here

$$\mathbf{C} = (c_1, c_2, \ldots, c_n), \qquad \mathbf{X} = \begin{pmatrix} x_1 \\ x_2 \\ \vdots \\ x_n \end{pmatrix},$$

$$\mathbf{A} = (a_{ij})_{m \times n}, \qquad \mathbf{b} = \begin{pmatrix} b_1 \\ b_2 \\ \vdots \\ b_m \end{pmatrix}.$$

There is an analogous minimization problem; that is,

$$\text{Minimize } \mathbf{CX}$$

subject to

$$\mathbf{AX} \geq \mathbf{b},$$
$$\mathbf{X} \geq \mathbf{0}.$$

We now list some features of the linear-programming problem:

(1) The function Z and each of the constraints is a linear function. This means that for each activity represented by x_i, its contribution to Z and to the drain on resources of each constraint is linearly proportional to the value of x_i. Also the total value of Z and the drain on each resource can be found by summing the contributions of the individual activities.
(2) The decision variables are assumed to be nonnegative. When this assumption is invalid we use a mathematical device to convert the model into one, with the same set of feasible solutions, for which $\mathbf{X} \geq 0$.
(3) The decision variables are assumed to be nonnegative real numbers. When this assumption is invalid (e.g., if some of the decision variables must be integers, an *integer-programming problem* results. (Methods for solution of this problem are given in Chapter 2.)
(4) Each constraint is assumed to be an inequality involving a $\leq$ or $\geq$ sign.

A constraint which is an equation

$$\sum_{i=1}^{m} a_{ij} x_i = b_j,$$

for some j, can be replaced by two inequalities in the constraint set

$$\sum_{i=1}^{m} a_{ij} x_i \leq b_i$$

and

$$\sum_{i=1}^{m} a_{ij}x_i \geq b_j.$$

Constraints which are strict inequalities (with $<$ or $>$ signs) are inadmissible.

We now define a "standard form for the linear-programming problem and then develop a method which will solve this standard form.

1.1.3. The LP Standard Form

$$\text{Maximize } \mathbf{CX} \tag{1.5}$$

subject to

$$\mathbf{AX} = \mathbf{b} \quad \text{(the constraints)}, \tag{1.6}$$

where

$$\mathbf{X} \geq \mathbf{0} \quad \text{(the nonnegativity conditions)}, \tag{1.7}$$

$$\mathbf{b} \geq \mathbf{0}.$$

The features of the standard form are:

(5) The objective is one of maximization.
(6) The constraints are all equations.
(7) The decision variables must be nonnegative.
(8) The constant, b_j, in each constraint is nonnegative.

We now show how to convert any LP problem, which we assume has the linear-programming features (1)–(4) into a problem with the standard-form features (5)–(8), preserving the same set of feasible solutions.

Steps

(i) If the objective is one of minimization, the objective function Z is multiplied by -1. Otherwise nothing is done to Z.
(ii) Each variable not constrained to be nonnegative is replaced everywhere in the formulation by the difference between two new variables, each of which is constrained to be nonnegative.
(iii) Each constraint with a negative constant is multiplied by -1.
(iv) Each constraint which is an inequality is converted into an equation by the addition of a new variable to the left-hand side (right-hand side) if it is a $\leq$ ($\geq$) constraint. These new variables are called *slack variables*. (The original variables are called *structural variables*.)

We now illustrate these rules by converting the following problem into standard form:

$$\text{Minimize } 4x_1 + 6x_2 \quad (= Z)$$

subject to

$$4x_1 + 12x_2 \leq -3,$$
$$3x_1 + 6x_2 = 4,$$

$$x_1 \geq 0, \qquad x_2 \text{ any real number.}$$

We begin by replacing the second constraint by the pair of inequalities

$$3x_1 + 6x_2 \leq 4 \quad \text{and} \quad 3x_1 + 6x_2 \geq 4,$$

so that (4) is obeyed. The steps are:

(i) As the objective is one of minimization, the Z is multiplied by -1.

(ii) x_2 is not constrained to be nonnegative. We therefore introduce two new variables x_3 and x_4 and set

$$x_2 = x_3 - x_4$$

and add

$$x_3 \geq 0 \quad \text{and} \quad x_4 \geq 0$$

to the nonnegativity conditions. x_2 is replaced by $x_3 - x_4$ throughout the formulation.

(iii) The first constraint is multiplied by -1:

$$-4x_1 - 12(x_3 - x_4) \geq 3.$$

This of course reverses the sense of the inequality.

(iv) The three inequalities

$$3x_1 + 6(x_3 - x_4) \leq 4,$$
$$3x_1 + 6(x_3 - x_4) \geq 4,$$
$$-4x_1 - 12(x_3 - x_4) \geq 3$$

are converted into equations by the introduction of the new slack variables x_5, x_6, and x_7:

$$3x_1 + 6(x_3 - x_4) + x_5 = 4,$$
$$3x_1 + 6(x_3 - x_4) = 4 + x_6,$$
$$-4x_1 - 12(x_3 - x_4) = 3 + x_7.$$

The problem in standard form is

$$\text{Maximize} \quad -4x_1 - 6x_3 + 6x_4$$

subject to

$$\begin{aligned} 3x_1 + 6x_3 - 6x_4 + x_5 &= 4, \\ 3x_1 + 6x_3 - 6x_4 - x_6 &= 4, \\ -4x_1 - 12x_3 + 12x_4 - x_7 &= 3, \\ x_1, x_3, x_4, x_5, x_6, x_7 &\geq 0. \end{aligned}$$

When the solution has been found, the equation

$$x_2 = x_3 - x_4$$

can be used to deduce the value of x_2.

The reader may be puzzled as to why we first converted the second constraint in a pair of inequalities only to convert each back to an equation. As standard-form feature (6) is satisfied for the constraint, it appears that these steps are redundant. Indeed some LP computer codes would not perform them. We do this double conversion here because it is very useful to introduce a slack variable into each constraint. This makes it far easier to deduce the behavior of the optimal solution and its value for a problem when changes are made to the data of the problem.

We now build up the basic concepts necessary to understand the method which solves problems in standard form.

1.1.4. Background

Consider the problem (1.5)–(1.7), where $\mathbf{X} = (x_1, x_2, \ldots, x_n)^T$ and $\mathbf{B} = (b_1, b_2, \ldots, b_m)^T$. A solution $\mathbf{X}$ is said to be *feasible* if it satisfies (1.6) and (1.7). Consider Eq. (1.6):

$$\mathbf{AX} = \mathbf{b}.$$

It is a system of m linear equations in n unknowns. Because a slack variable has been introduced into each constraint we have

$$m < n.$$

Suppose a set X_N, of $n - m$ variables out of $x_1, x_2, \ldots, x_n$, is set equal to zero. Suppose further that a submatrix $\mathbf{B}$ is formed from $\mathbf{A}$ by deleting column j if x_j is set to zero. Thus $\mathbf{B}$ is an $m \times m$ submatrix of $\mathbf{A}$. Let the remaining variables in $\mathbf{X}$, not set to zero, be denoted by $\mathbf{X_B}$. We have reduced (1.6) to

$$\mathbf{BX_B} = \mathbf{b}.$$

If **B** is nonsingular (see Section 5.1 for an explanation of this term) we can solve for $\mathbf{X_B}$:

$$\mathbf{X_B} = \mathbf{B}^{-1}\mathbf{b}. \tag{1.8}$$

Recall that

$$\mathbf{X_N} = \mathbf{0}. \tag{1.9}$$

Together (1.8) and (1.9) comprise a solution to (1.6). Such a solution is called a *basic solution*. The variables in $\mathbf{X_B}$ are termed *basic variables* and the complete collection of basic variables is called a *basis*. The variables in $\mathbf{X_N}$ are *nonbasic variables*. If a basic solution also satisfies (1.7) it is called a *basic feasible solution*. A basic feasible solution is termed *degenerate* if at least one of the basic variables has value zero.

The reader may have noticed that the feasible region in Fig. 1.1 has the following property. The straight line joining any two points in the region lies entirely within the region. Any set with this property is termed "convex." This definition can be extended to sets in higher dimensions in a natural way. Using the analogous definition for higher dimensions, it can be shown that the set of feasible solutions to any LP problem is convex. As the objective function is linear, it is clear that the optimal solution to the problem shown in Fig. 1.1 must lie at one of the "corners" of the convex feasible region. Such points, which cannot be expressed as a linear combination of other points in the region, are called *extreme points*. Thus we need only examine the extreme points of the feasible region in order to find an optimal solution (if such a solution exists). The *simplex method* examines the extreme points in an efficient way in order to find an optimal solution. We shall introduce it in the next section.

1.1.5. The Simplex Method

1.1.5.1. *Canonical Form.* It can be shown that there is a one-to-one correspondence between the basic solutions of an LP problem and the extreme points of its feasible region. Thus, bearing in mind the last remark of Section 1.1.4, we need to examine only the basic solutions of a problem in order to be certain of finding an optimal solution. As an optimal solution is, by definition, feasible we need to examine only the basic feasible solutions. The simplex method generates these solutions one at a time, without decreasing Z in going from one solution to the next. It does this in such a way that when no further increase in Z is possible the optimal solution has been found. The change in the basic feasible solution is achieved by making one of the basic variables nonbasic and one of the nonbasic variables basic. This is done by setting one of the basic variables to zero and selecting a new variable to be basic.

Equation (1.8) is solved to obtain values for all the basic variables, including the newcomer. The incoming variable is selected in order to make the largest possible unit increase in Z. The variable departing from $\mathbf{X_B}$ is selected on the criterion of ensuring that the new basic solution remains feasible.

We now introduce a new illustrative example:

$$\text{Maximize } 6x_1 + 4x_2 + 3x_3$$

subject to

$$\begin{aligned} 4x_1 + 5x_2 + 3x_3 &\leq 12, \\ 3x_1 + 4x_2 + 2x_3 &\leq 10, \\ 4x_1 + 2x_2 + x_3 &\leq 8, \\ x_1, x_2, x_3 &\geq 0. \end{aligned}$$

In order to convert the problem into standard form, we introduce slack variables x_4, x_5, and x_6:

$$\text{Maximize } 6x_1 + 4x_2 + 3x_3$$

subject to

$$4x_1 + 5x_2 + 3x_3 + x_4 = 12, \tag{1.10}$$

$$3x_1 + 4x_2 + 2x_3 + x_5 = 10, \tag{1.11}$$

$$4x_1 + 2x_2 + x_3 + x_6 = 8, \tag{1.12}$$

$$x_1, x_2, \ldots, x_6 \geq 0.$$

Let

$$\mathbf{X_B} = (x_4, x_5, x_6)^T.$$

$$\therefore \quad \mathbf{X_N} = (x_1, x_2, x_3)^T.$$

and

$$\mathbf{B} = \mathbf{I}.$$

Also,

$$\mathbf{XB} = \mathbf{b}.$$

$$\therefore \quad \mathbf{X} = \mathbf{B}^{-1}\mathbf{b}$$

and

$$\mathbf{X} = \mathbf{b} = (12, 10, 8)^T.$$

As all the variables are nonnegative, we have a basic feasible solution. Its value is

$$\begin{aligned} Z &= (6,4,3)(0,0,0)^T \\ &= 0. \end{aligned}$$

Let us find a new basic feasible solution with a larger value. We wish to replace one of the basic variables by a nonbasic variable. As x_1 has the largest coefficient in Z let us choose it. We must decide which variables out of x_4, x_5, and x_6 must depart from the basis. Two considerations guide us: (i) we want to make x_1 as large as possible in order to increase Z as much as possible; and (ii) the new basis must be such that all of its variables are nonnegative. In order to settle this we solve each constraint for the incoming variable x_1:

$$x_1 = 3 - \tfrac{5}{4}x_2 - \tfrac{3}{4}x_3 - \tfrac{1}{4}x_4,$$

$$x_1 = \tfrac{10}{3} - \tfrac{4}{3}x_2 - \tfrac{2}{3}x_3 - \tfrac{1}{3}x_5,$$

$$x_1 = 2 - \tfrac{1}{2}x_2 - \tfrac{1}{4}x_3 - \tfrac{1}{4}x_6.$$

As x_2 and x_3 are nonbasic, they are zero. So these equations reduce to

$$x_1 = 3 - \tfrac{1}{4}x_4, \tag{1.10'}$$

$$x_1 = \tfrac{10}{3} - \tfrac{1}{3}x_5, \tag{1.11'}$$

$$x_1 = 2 - \tfrac{1}{4}x_6. \tag{1.12'}$$

Let us consider the removal of each of the x_4, x_5, and x_6 in turn from the basis. This entails setting each of them equal to zero:

$$x_4 = 0 \Rightarrow x_1 = 3 \qquad [\text{from } (1.10')],$$

$$x_5 = 0 \Rightarrow x_1 = \tfrac{10}{3} \qquad [\text{from } (1.11')],$$

$$x_6 = 0 \Rightarrow x_1 = 2 \qquad [\text{from } (1.12')].$$

But if $x_1 = 3$ or $\frac{10}{3}$ then by (1.12′), $x_6 < 0$ which is infeasible. Thus x_4 or x_6 cannot leave the basis.

However, if $x_1 = 2$, x_4 and $x_5 > 0$ and so a feasible basis is maintained. So the only variable which can safely leave the basis is x_6. There is a quick way to establish which variable departs—identify the minimum positive constant in equations of the form (1.10)–(1.12). The basic variable in the equation with this minimum positive constant departs.

$$\mathbf{X_B} = (x_1, x_4, x_5)^T, \qquad \mathbf{BX_B} = \mathbf{b}$$

becomes

$$\begin{pmatrix} 4 & 1 & 0 \\ 3 & 0 & 1 \\ 4 & 0 & 0 \end{pmatrix} \begin{pmatrix} x_1 \\ x_4 \\ x_5 \end{pmatrix} = \begin{pmatrix} 12 \\ 10 \\ 8 \end{pmatrix}$$

because

$$\mathbf{A} = \begin{pmatrix} 4 & 5 & 3 & 1 & 0 & 0 \\ 3 & 4 & 2 & 0 & 1 & 0 \\ 4 & 2 & 1 & 0 & 0 & 1 \end{pmatrix}$$

and we are now defining $\mathbf{B}$ to be made up of columns, 1, 4, and 5 corresponding to x_1, x_4, and x_5.

$$\begin{pmatrix} x_1 \\ x_4 \\ x_5 \end{pmatrix} = \begin{pmatrix} 4 & 1 & 0 \\ 3 & 0 & 1 \\ 4 & 0 & 0 \end{pmatrix}^{-1} \begin{pmatrix} 12 \\ 10 \\ 8 \end{pmatrix} = \begin{pmatrix} 2 \\ 4 \\ 4 \end{pmatrix}$$

and

$$\begin{aligned} Z &= (6,4,3)(2,0,0)^T \\ &= 12. \end{aligned}$$

We have, in essence, performed one iteration of the simplex method. We now show how to transform the solution to optimality.

The calculations of the simplex method are now facilitated by arranging the initial data in a special table (Table 1.1). Note that each row corresponds to a constraint except the last row, which corresponds to $-Z$. The last entry in the $-Z$ row is the current value of the objective function. Each column corresponds to a variable except the last two, which correspond to b and the ratios formed in (1.10′)–(1.12′).

Table 1.1

	x_1	x_2	x_3	x_4	x_5	x_6	b	Ratios	
Eq. (1.10)	4	5	3	1	0	0	12	$\frac{12}{4}$	R_1
Eq. (1.11)	3	4	2	0	1	0	10	$\frac{10}{3}$	R_2
Eq. (1.12)	④	2	1	0	0	1	8	$\frac{8}{4}$	R_3
$-Z$	-6	-4	-3	0	0	0	0		R_4

The simplex method begins by selecting an initial basis, corresponding to a basic feasible solution. In our case an obvious basis is the set of slack variables: x_4, x_5, and x_6. As we shall see, such a selection does not always correspond to a feasible solution. Special techniques which overcome this problem are dealt with later. In any table produced by the simplex method each basic variable has a value equal to the b value in its row. Of course, each nonbasic variable has a value of 0.

As can be seen from looking at Table 1.1:

(i) **B** forms an identity matrix within a permutation of columns.
(ii) The entries in the $-Z$ row corresponding to basic variables are zero. A table for which (i) and (ii) hold is said to be in *canonical form*. The simplex method produces a series of tables in canonical form, each corresponding to an improved basic feasible solution.

Recall that we have decided to replace x_6 by x_1 in the basis. We must now create the new table, in canonical form, with basis x_1, x_4, and x_5 satisfying (i) and (ii). This means that the x_1 column must be transformed until it looks like the present x_6 column: $(0,0,1,0)^T$. It can be shown that

Gauss–Jordon elimination (see Section 5.1) can be used to achieve this without affecting the set of feasible solutions to the original problem.

The entry in the table which is at the intersection of the row containing the minimum ratio and the column of the entering variable is called the *pivot*. It is represented by a solid circle in Table 1.1.

The Gauss–Jordon elimination is now used to produce an x_1 column of $(0,0,1,0)^T$ and hence a table satisfying (i) and (ii). It is shown in Table 1.2. The row operations used are shown to the right. The row labels are given in Table 1.1.

Table 1.2

	x_1	x_2	x_3	x_4	x_5	x_6	b	Ratios	
Eq. (1.10)	0	3	②	1	0	-1	4	$\frac{4}{12}$	$R_1 - R_3$
Eq. (1.11)	0	$\frac{5}{2}$	$\frac{5}{4}$	0	1	$-\frac{3}{4}$	4	$\frac{16}{5}$	$R_2 - \frac{3}{4}R_3$
Eq. (1.12)	1	$\frac{1}{2}$	$\frac{1}{4}$	0	0	$\frac{1}{4}$	2	8	$\frac{1}{4}R_3$
$-Z$	0	-1	$-\frac{3}{2}$	0	0	$\frac{3}{2}$	12		$R_4 + \frac{6}{4}R_3$

We have now produced the second basic feasible solution. As there are still negative entries in the $-Z$ row, Table 1.2 does not represent an optimal solution.

Thus we can bring at least one of the entries into the basis (with positive value) and increase Z. (Recall that we are maximizing and that a minimization problem is converted to one of maximization by step (i) of Section 1.1.3.

We have to find the nonbasic variable with the most negative entry in the $-Z$ row for entry into the basis. This is x_3. On taking ratios, we see that the minimum occurs in the top row. Thus x_4 departs. Performing the necessary transformation to ensure that the new table obeys steps (ii) and (iii) of Section 1.1.3 produces Table 1.3.

Table 1.3

	x_1	x_2	x_3	x_4	x_5	x_6	b
Eq. (1.10)	0	$\frac{3}{2}$	1	$\frac{1}{2}$	0	$-\frac{1}{2}$	2
Eq. (1.11)	0	$\frac{5}{8}$	0	$-\frac{5}{8}$	1	$-\frac{1}{8}$	$\frac{3}{2}$
Eq. (1.12)	1	$\frac{1}{8}$	0	$-\frac{1}{8}$	0	$\frac{3}{8}$	$\frac{3}{2}$
$-Z$	0	$\frac{5}{4}$	0	$\frac{3}{4}$	0	$\frac{3}{4}$	15

As there are no more negative entries in the $-Z$ row, this table represents the optimal solution

$$x_1^* = \tfrac{3}{2},$$

$$x_3^* = 2,$$

$$x_5^* = \tfrac{3}{2},$$

$$x_2^* = x^* = x_6^* = 0,$$

$$Z^* = 15.$$

1.1.5.2. *Artificial Variables*. In the example problems introduced in the previous sections all the constraints have been inequalities of the $\leq$ type. This has allowed us to add a slack variable to the left-hand side of each constraint in order to convert it into an equation. The **B** matrix corresponding to a basis of all slack variables is the identity matrix. Thus

$$\mathbf{X} = \mathbf{B}^{-1}\mathbf{b} = \mathbf{b}$$

is a basic feasible solution as

$$\mathbf{b} > 0$$

and $\mathbf{B}^{-1}$ is nonsingular. It is usual to define **B** in this way as it provides an initial basic feasible solution which is easy to find.

However, if not all of the constraints are of the $\leq$ type, the initial basis of slack variables is not feasible. In this case we must introduce extra variables, called *artificial variables*, in order to find an initial basic feasible solution. We now illustrate this point with a numerical example:

$$\text{Maximize } 4x_1 + 5x_2 \quad (= Z)$$

subject to

$$3x_1 + 5x_2 \leq 24,$$

$$4x_1 + 2x_2 \leq 16,$$

$$x_1 + x_2 \geq 3,$$

$$x_1, \ x_2 \geq 0.$$

We transform the problem into standard form by introducing slack variables x_3, x_4, and x_5:

$$\text{Maximize } 4x_1 + 5x_2 \qquad\qquad (= Z)$$

subject to

$$\begin{aligned} 3x_1 + 5x_2 + x_3 \qquad\qquad &= 24, \\ 4x_1 + 2x_2 \qquad + x_4 \qquad &= 16, \\ x_1 + x_2 \qquad\qquad - x_5 &= 3, \\ x_1, x_2, \ldots, x_5 &\geq 0. \end{aligned}$$

If we let the slack variables form the initial basis, we have

$$\mathbf{B} = \begin{pmatrix} 1 & 0 & 0 \\ 0 & 1 & 0 \\ 0 & 0 & -1 \end{pmatrix}$$

and

$$\mathbf{X} = \mathbf{B}^{-1}\mathbf{b}$$

$$= \begin{pmatrix} 24 \\ 16 \\ -3 \end{pmatrix},$$

which is infeasible as $x_5 < 0$.

This dilemma is overcome by introducing an additional variable into any row which does not have an added slack variable. In the present case this means adding the nonnegative *artificial* variable x_6 to the left-hand side of the last constraint. The problem becomes:

$$\text{Maximize } 4x_1 + 5x_2 \qquad (=Z)$$

subject to

$$3x_1 + 5x_2 + x_3 = 24, \qquad (1.13)$$

$$4x_1 + 2x_2 + x_4 = 16, \qquad (1.14)$$

$$x_1 + x_2 - x_5 + x_6 = 3, \qquad (1.15)$$

$$x_1, x_2, \ldots, x_6 \geq 0.$$

The **A** matrix is now

$$\begin{matrix} & x_1 & x_2 & x_3 & x_4 & x_5 & x_6 \end{matrix}$$

$$\mathbf{A} = \begin{pmatrix} 3 & 5 & 1 & 0 & 0 & 0 \\ 4 & 2 & 0 & 1 & 0 & 0 \\ 1 & 1 & 0 & 0 & -1 & 1 \end{pmatrix}.$$

If we take as our initial basis x_3, x_4, and x_6,

$$\mathbf{B} = \mathbf{I},$$

which corresponds to a basic feasible solution. In general, the initial basis is formed by taking (i) all the artificial variables and (ii) the slack variables from the constraints which do not have artificial variables.

Of course we cannot simply add a variable to one side of an equation and expect the equation to be satisfied for any values other than zero for the new variable. We know that in the optimal solution

$$x_6^* = 0.$$

We must modify the version of the simplex method previously outlined to ensure that it eliminates all artificial variables from the basis. One technique for doing this is called the *two-phase method*.

1.1.5.3. *The Two-Phase Method*. Phase I of the method begins by replacing the objective function by

$$\text{Minimize } \sum x_j,$$

where each x_j in the above summation is an artificial variable. Thus a new objective is defined which is to minimize the sum of the artificial variables subject to the original set of constraints and nonnegativity conditions. In our case the only artificial variable is x_6 so the new LP problem is

$$\text{Minimize } x_6 \qquad (=Z') \qquad (1.16)$$

subject to

$$\begin{aligned} 3x_1 + 5x_2 + x_3 \qquad\qquad &= 24,\\ 4x_1 + 2x_2 \qquad + x_4 \qquad &= 16,\\ x_1 + x_2 \qquad\qquad - x_5 + x_6 &= 3,\\ x_1, x_2, \ldots, x_6 &\geq 0. \end{aligned}$$

We now apply step (i) of Section 1.1.3 to make the objective one of maximization:

$$\text{Maximize } -x_6 \quad (=Z'')$$

The initial table for the new problem is shown as Table 1.4. This table is not in canonical form as the bottom-row entries for the basis, x_3, x_4, and x_6, are not all zero (the entry for x_6 is 1). We can subtract (1.15) from $(-Z'')$ to transform the table into canonical form as shown in Table 1.5. We now use the simplex method to complete the first phase. Either x_1 and x_2 can enter the basis at this stage. We choose x_2 arbitrarily. This produces Table 1.6. As all the $-Z''$ row entries are nonnegative, we have solved the new LP problem, Eqs. (1.13)–(1.16). The optimal solution is

$$x_2^* = 3, \qquad x_3^* = 9, \qquad x_4^* = 10, \quad \text{and} \quad Z'' = 0.$$

Table 1.4

	x_1	x_2	x_3	x_4	x_5	x_6	b
Eq. (1.13)	3	5	1	0	0	0	24
Eq. (1.14)	4	2	0	1	0	0	16
Eq. (1.15)	1	1	0	0	−1	1	3
$-Z''$	0	0	0	0	0	1	0

Table 1.5

	x_1	x_2	x_3	x_4	x_5	x_6	b	Ratios
Eq. (1.13)	3	5	1	0	0	0	24	$\frac{24}{5}$
Eq. (1.14)	4	2	0	1	0	0	16	$\frac{16}{2}$
Eq. (1.15)	1	1	0	0	−1	1	3	$\frac{3}{1}$
$-Z''$	−1	−1	0	0	1	0	−3	

Table 1.6

	x_1	x_2	x_3	x_4	x_5	x_6	b
Eq. (1.13)	-2	0	1	0	5	-5	9
Eq. (1.14)	2	0	0	1	2	-2	10
Eq. (1.15)	1	1	0	0	-1	1	3
$-Z''$	0	0	0	0	0	1	0

Table 1.7

	x_1	x_2	x_3	x_4	x_5	b
Eq. (1.13)	-2	0	1	0	5	9
Eq. (1.14)	2	0	0	1	2	10
Eq. (1.15)	1	1	0	0	-1	3
$-Z$	-4	-5	0	0	0	0

Table 1.8

	x_1	x_2	x_3	x_4	x_5	b	Ratios
Eq. (1.13)	-2	0	1	0	5	9	$\frac{9}{5}$
Eq. (1.14)	2	0	0	1	2	10	$\frac{10}{12}$
Eq. (1.15)	1	1	0	0	-1	3	
$-Z$	1	0	0	0	-5	15	

Table 1.9

	x_1	x_2	x_3	x_4	x_5	b	Ratios
Eq. (1.13)	$-\frac{2}{5}$	0	$\frac{1}{5}$	0	1	$\frac{9}{5}$	
Eq. (1.14)	$\frac{14}{5}$	0	$-\frac{2}{5}$	1	0	$\frac{32}{5}$	$\frac{32}{14}$
Eq. (1.15)	$\frac{3}{5}$	1	$\frac{1}{5}$	0	0	$\frac{34}{5}$	$\frac{24}{3}$
$-Z$	-1	0	1	0	0	24	

Table 1.10

	x_1	x_2	x_3	x_4	x_5	b
Eq. (1.13)	0	0	$\frac{1}{7}$	$\frac{1}{7}$	1	$\frac{19}{7}$
Eq. (1.14)	1	0	$-\frac{1}{7}$	$\frac{5}{14}$	0	$\frac{16}{7}$
Eq. (1.15)	0	1	$\frac{2}{7}$	$-\frac{3}{14}$	0	$\frac{24}{7}$
$-Z$	0	0	$\frac{6}{7}$	$\frac{5}{14}$	0	$\frac{164}{7}$

We have succeeded in eliminating x_6 from the basis and thus $Z''=0$. This solution represents a feasible basis for the original problem. This is what we were looking for so now we can return to the original problem. x_6 is deleted from the initial table and the original objective function row $-Z$ replaces $-Z''$. This is shown in Table 1.7. We now begin the second phase.

Note that the entry in the bottom right-hand corner is set at zero. It reflects the true value of Z when the table is converted into standard form. This is done in Table 1.8.

The simplex method is now used to solve the original problem, starting from Table 1.7. Note that one of the ratios is not formed in the table as its denominator is not positive. The calculations are shown in Tables 1.9 and 1.10. Table (1.10) represents the optimal solution:

$$x_1^* = \tfrac{16}{7}, \quad x_2^* = \tfrac{24}{7}, \quad x_5^* = \tfrac{19}{7}, \text{ and } Z^* = \tfrac{164}{7}.$$

Summary of the Two-Phase Method

It is assumed that the LP problem is in standard form.

Phase I

(1) Add an artificial variable to the left-hand side of each constraint which has a subtracted slack variable.
(2) Replace the objective function by the minimization of the sum of the artificial variables (denoted by Z').
(3) Transform the initial table into canonical form.
(4) Apply the simplex method to this table.
  (i) If $(Z')^* > 0$, the original problem does not have any feasible solutions and the method is terminated.
  (ii) If $(Z')^* = 0$, apply phase II.

Phase II

(5) Remove all artificial-variable columns from the final tableau produced in step (4). Replace the $-Z'$ row by $-Z$, where Z is the original objective function.
(6) Transform the table produced in step (5) into canonical form.
(7) Apply the simplex method to this table.

1.1.5.4. *Summary of the Simplex Method.* It is assumed that the LP problem is in standard form.

Steps

(1) Set up the initial table.

Criterion for Optimality

(2) Find the most negative entry in the $-Z$ row. If all are nonnegative, the table represents an optimal solution and the method is terminated. Otherwise, label the column of the entry as j. (Ties are settled arbitrarily.) Go to step (3).
(3) For each positive entry a'_{ij} in column j, form the ratio b'_i/a'_{ij} where b'_i is the entry in the b column in row i.
(4) Choose the minimum ratio of those formed in step (3), say b'_l/a'_{lk}. (Ties are settled arbitrarily.)
(5) Replace x_i in the basis by x_k by the Gauss–Jordon elimination.
(6) Go back to step (2).

1.1.5.5. *Multiple Optima.* Consider the following simple LP problem in standard form:

$$\text{Maximize } 4x_1 + 5x_2 \quad (= Z) \tag{1.17}$$

subject to

$$4x_1 + 5x_2 + x_3 = 32 \tag{1.18}$$

$$4x_1 + 3x_2 + x_4 = 24, \tag{1.19}$$

$$x_1, x_2 \geq 0$$

The problem is solved by the simplex method in Tables 1.11–1.13. Table (1.12) represents an optimal solution as the criterion for optimality of step (2) of Section 1.1.5.4 is satisfied. The optimal solution is

$$x_1^* = 0, \quad x_2^* = \tfrac{32}{5}, \quad x_3^* = 0, \quad x_4^* = \tfrac{24}{5}, \quad \text{and} \quad Z = 32.$$

However, it can be seen that the $-Z$ entry of the x_1 column is zero. Bringing x_1 into the basis is achieved in Table 1.13. Table 1.13 represents

Table 1.11

	x_1	x_2	x_3	x_4	b	Ratios
Eq. (1.18)	4	5	1	0	32	$\frac{32}{5}$
Eq. (1.19)	4	3	0	1	24	$\frac{24}{3}$
$-Z$	-4	-5	0	0	0	

Table 1.12

	x_1	x_2	x_3	x_4	b	Ratios
Eq. (1.18)	$\frac{4}{5}$	1	$\frac{1}{5}$	0	$\frac{32}{5}$	$\frac{32}{4}$
Eq. (1.19)	$\frac{8}{5}$	0	$-\frac{3}{5}$	1	$\frac{24}{5}$	$\frac{24}{8}$
$-Z$	0	0	1	0	32	

Table 1.13

	x_1	x_2	x_3	x_4	b
Eq. (1.18)	0	1	$\frac{1}{2}$	$-\frac{1}{2}$	4
Eq. (1.19)	1	0	$-\frac{3}{8}$	$\frac{5}{8}$	3
$-Z$	0	0	1	0	32

another optimal solution:

$$x_1^* = 3, \quad x_2^* = 4, \quad x_3^* = 0, \quad x_4^* = 0, \quad \text{and} \quad Z^* = 32.$$

Of course, as the $-Z$ row coefficient of x_4 is zero we can bring it into the basis, returning to Table 1.12. We have discovered two distinct optimal basic solutions. In order to discover how this can come about, please refer to Fig. 1.2 which depicts the feasible region.

The objective function Z has the same slope as (1.18). It means that in this case any point on the line segment of (1.19) from $(0, \frac{32}{5})$ to $(3,4)$ represents an optimal solution. We can state this as follows:

$$4x_1^* + 5x_2^* = 32,$$

$$0 \leq x_1^* \leq 3,$$

$$4 \leq x_2^* \leq \tfrac{32}{5},$$

$$Z^* = 32.$$

Multiple optima can be detected in the final simplex tableau by the presence of a nonbasic column having a zero entry in the $-Z$ *row.*

1.1.5.6. *Degeneracy.* The simplex method is based on the assumption that each new basic feasible solution value is an improvement over its predecessor. As we assume maximization this means that Z must strictly increase from one basis to the next. However, suppose that one of the basic variables has value zero in an intermediate simplex table (i.e., one of the entries in the column is zero). A basis with this property is termed "*degenerate.*" When ratios are formed, the zero will be the numerator of the minimum ratio. When the change of basis is made, Z will not have increased. This can possibly lead to a series of tables, each with the same value of Z. Even worse, there is

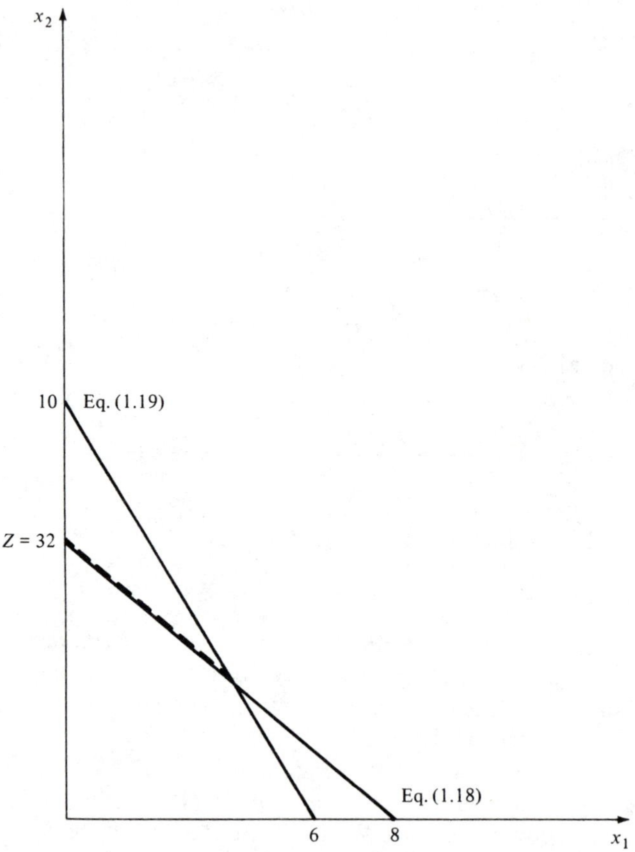

Figure 1.2. Multiple Optima.

the possibility that one of the earlier bases will reappear, leading to an endless series of tables, all with suboptimal solutions. This is called *cycling*.

Degeneracy often occurs in large-scale real-world problems. However, cycling in such instances is very rare. There exist techniques to prevent cycling, but they are seldom used in commercial LP computer codes as the accumulation of errors usually prevents cycling.

1.1.5.7. *No Feasible Solution*. Consider the following LP problem in standard form in which the artificial variable x_6 has been introduced:

$$\text{Maximize } 4x_1 + 3x_2 \qquad (= Z)$$

subject to

$$3x_1 + 4x_2 + x_3 = 12, \tag{1.20}$$

$$5x_1 + 2x_2 + x_4 = 8, \tag{1.21}$$

$$x_1 + x_2 - x_5 + x_6 = 5, \tag{1.22}$$

$$x_1, x_2, \ldots, x_6 \geq 0.$$

We now apply the two-phase simplex method to the problem as shown in Tables 1.14–1.17.

Table 1.14

	x_1	x_2	x_3	x_4	x_5	x_6	b
Eq. (1.20)	3	4	1	0	0	0	12
Eq. (1.21)	5	2	0	1	0	0	8
Eq. (1.22)	1	1	0	0	-1	1	5
$-Z''$	0	0	0	0	0	1	0

Table 1.15

	x_1	x_2	x_3	x_4	x_5	x_6	b
Eq. (1.20)	3	4	1	0	0	0	12
Eq. (1.21)	5	2	0	1	0	0	8
Eq. (1.22)	1	1	0	0	-1	1	5
$-Z''$	-1	-1	0	0	-1	0	-5

Table 1.16

	x_1	x_2	x_3	x_4	x_5	x_6	b
Eq. (1.20)	$\frac{3}{4}$	1	$\frac{1}{4}$	0	0	0	3
Eq. (1.21)	$\frac{7}{2}$	0	$-\frac{1}{2}$	1	0	0	2
Eq. (1.22)	$\frac{1}{4}$	0	$-\frac{1}{4}$	0	-1	1	2
$-Z''$	$-\frac{1}{4}$	0	$\frac{1}{4}$	0	-1	0	-2

Table 1.17

	x_1	x_2	x_3	x_4	x_5	x_6	b
Eq. (1.20)	0	1	$\frac{5}{14}$	$-\frac{3}{14}$	0	0	$\frac{18}{7}$
Eq. (1.21)	1	0	$-\frac{1}{7}$	$\frac{2}{7}$	0	0	$\frac{4}{7}$
Eq. (1.22)	0	0	$-\frac{3}{14}$	$-\frac{1}{14}$	-1	1	$\frac{13}{7}$
$-Z''$	0	0	$\frac{3}{14}$	$\frac{1}{14}$	1	0	$-\frac{13}{7}$

Phase I

We have reached the final table in Table 1.17 as the criterion for optimality [step (2) in Section 1.1.5.4] is satisfied. And yet the artificial variable is still

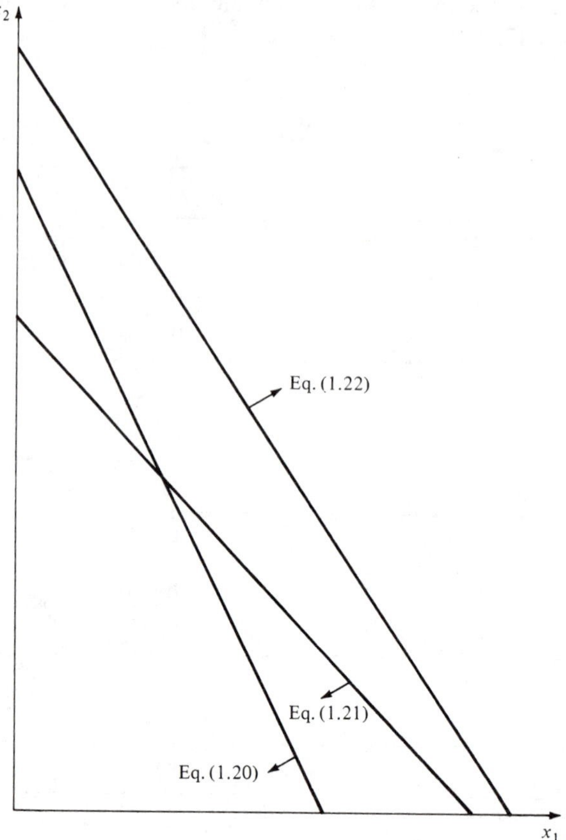

Figure 1.3. No Feasible Solutions.

basic and indeed positive:

$$x_6^* = \tfrac{13}{7} > 0.$$

The first phase has failed to eliminate the artificial variable from the basis. This means that the original problem has no feasible solution. In order to see why this is so, please study Fig. 1.3.

As can be seen a point does not exist which satisfies all the constraints and nonnegativity conditions simultaneously. Hence there is no feasible solution to the problem. The two-phase method detects infeasibility by having at least one artificial variable still positive at the end of the first phase. In this case phase II is not attempted.

1.1.6. Duality

One of the most important concepts in linear programming is that of *duality*. Apart from having theoretical interest in its own right, duality is used to solve LP problems more efficiently than by simply applying the simplex method to them. It is also used to answer questions concerning changes to the data of a problem.

Consider the following LP problem P_1:

$$\text{Maximize } 5x_1 + 4x_2 \qquad (=Z)$$

subject to

$$3x_1 + 4x_2 \le 14, \tag{1.23}$$

$$4x_1 + 2x_2 \le 8, \tag{1.24}$$

$$2x_1 + x_2 \le 6, \tag{1.25}$$

$$x_1, x_2 \ge 0.$$

We can define another closely related LP problem D_1:

$$\text{Minimize } 14y_1 + 8y_2 + 6y_3 \quad (=W) \tag{1.26}$$

subject to

$$3y_1 + 4y_2 + 2y_3 \ge 5, \tag{1.27}$$

$$4y_1 + 2y_2 + y_3 \ge 4, \tag{1.28}$$

$$y_1, y_2, y_3 \ge 0.$$

Let us compare the two problems. It is easily seen that they have the same constants, but in different positions. When a second LP problem has the relationship to another LP problem, shown in the example, it is called *dual.* The original problem is called the *primal.* We now formalize the relationship between the dual and the primal:

(1) One problem P_1 has a maximization objective, the other D_1, a minimization objective.
(2) Each constraint in P_1 is a $\le$ inequality. Each constraint in D_1 is a $\ge$ inequality.
(3) Each constraint in one problem corresponds to a nonnegative variable in the other problem.
(4) The objective function coefficient vector in one problem corresponds to the vector of constants in the constraints of the other problem.
(5) The matrix of constraint coefficients in one problem is the transpose of the matrix of constraint coefficients in the other problem.

These relationships can be summarized mathematically. The primal P

$$\text{Maximize } c_1x_1 + c_2x_2 + \cdots + c_nx_n \quad (=Z)$$

subject to

$$\begin{aligned}
&a_{11}x_1 + a_{12}x_2 + \cdots + a_{1n}x_n \le b_1,\\
&a_{21}x_1 + a_{22}x_2 + \cdots + a_{2n}x_n \le b_2,\\
&\vdots\\
&a_{m1}x_1 + a_{m2}x_2 + \cdots + a_{mn}x_n \le b_m,\\
&x_1, x_2, \ldots, x_n \ge 0,
\end{aligned}$$

has dual D,

$$\text{Minimize } b_1y_1 + b_2y_2 + \cdots + b_my_m \quad (=W)$$

subject to

$$\begin{aligned}
&a_{11}y_1 + a_{22}y_2 + \cdots + a_{m1}y_m \ge c_1,\\
&a_{12}y_1 + a_{22}y_2 + \cdots + a_{m2}y_m \ge c_2,\\
&\vdots\\
&a_{1n}y_1 + a_{2n}y_2 + \cdots + a_{mn}y_m \ge c_n,\\
&y_1, y_2, \ldots, y_m \ge 0.
\end{aligned}$$

In matrix form primal P

$$\text{Maximize } \mathbf{CX} \tag{1.29}$$

subject to

$$\mathbf{AX} \le \mathbf{b}, \tag{1.30}$$

$$\mathbf{X} \ge 0, \tag{1.31}$$

has dual D,

$$\text{Minimize } \mathbf{b}^T\mathbf{y}$$

subject to

$$\mathbf{A}^T\mathbf{y} \ge \mathbf{C}^T,$$

$$\mathbf{Y} \ge \mathbf{0},$$

$$\mathbf{Y} = (y_1, y_2, \ldots, y_m)^T.$$

Note that there are nonnegativity conditions in both problems.

We can use these relationships to devise a recipe for creating the dual D of any LP problem from a primal LP problem P:

(1) Replace each primal constraint by a pair of inequalities, one with sense $\leq$ and one with sense $\geq$. $\Sigma_{i=1}^{n} a_{ij}x_j = b_j$ becomes

$$\sum_{i=1}^{n} a_{ij}x_i \leq b_j$$

and

$$\sum_{i=1}^{n} a_{ij}x_i \geq b_j.$$

(Recall that this is also one of the rules for transforming an LP problem into standard form.)

(2) If P has a maximization objective, multiply all $\geq$ constraints by -1 to ensure that all constraints are of the $\leq$ type and define D to have a minimization objective with all constraints of the $\geq$ type. If P has a minimization objective, multiply all $\leq$ constraints by -1 to ensure that all constraints are of the $\geq$ type and define D to have a maximization objective with all constraints of the $\leq$ type.
(3) For each constraint in P define a unique structural variable for D.
(4) Define the objective function coefficient of each variable in D to be the constant of the constraint in P with which it is associated.
(5) Define the vector of constants of the constraints in D to be the transpose of the vector of the objective function coefficients in P.
(6) Define the matrix of constraint coefficients to be the transpose of the primal constraint matrix.

We now use these rules to create the dual D_1, of P_1:

(1) There is nothing to be done here.
(2) P_1 must have a maximization objective with all constraints of the $\geq$ type.
(3) and (4) Define y_1 for (1.23), with objective function coefficient 14; y_2 for (1.24), with objective function coefficient 8; and y_3 for (1.25), with objective function coefficient 6.
(5) As $C = (5,4)$ in P_1, define $b = (5,4)^T$ in D_1.
(6) As

$$\mathbf{A} = \begin{pmatrix} 3 & 4 \\ 4 & 2 \\ 2 & 1 \end{pmatrix} \qquad \text{for } P_1,$$

the corresponding matrix for D_1 is

$$\begin{pmatrix} 3 & 4 & 2 \\ 4 & 2 & 1 \end{pmatrix}.$$

Thus we obtain the dual problem (1.26)–(1.28). It is automatically assumed that a nonnegativity condition holds for each variable as well.

Let us take the dual of LP problem (1.29)–(1.31). Applying the above rules we obtain the problem:

$$\text{Maximize } (\mathbf{C}^T)^T\mathbf{X}$$

subject to

$$(\mathbf{A}^T)^T\mathbf{X} \leq (\mathbf{b}^T)^T,$$
$$\mathbf{X} \geq \mathbf{0}.$$

But as the transpose of the transpose of a vector (or matrix) is just the vector (or matrix) itself, we have shown that the dual of the dual of an LP problem is the LP problem itself.

Let us now solve D_1. First we introduce slack variables y_4 and y_6 and artificial variables y_5 and y_7. In standard form the problem is

$$\text{Maximize } -14y_1 - 8y_2 - 6y_3 \qquad (=W'')$$

subject to

$$3y_1 + 4y_2 + 2y_3 - y_4 + y_5 = 5$$
$$4y_1 + 2y_2 + y_3 - y_6 + y_7 = 4$$
$$y_1, y_2, \ldots, y_7 \geq 0.$$

The calculations necessary to find the optimal solution are given in Tables 1.18–1.23.

Phase I

Table 1.18

	y_1	y_2	y_3	y_4	y_5	y_6	y_7	b
Eq. (1.27)	3	4	2	−1	1	0	0	5
Eq. (1.28)	4	2	1	0	0	−1	1	4
$-W''$	0	0	0	0	1	0	1	0

Table 1.19

	y_1	y_2	y_3	y_4	y_5	y_6	y_7	b	Ratios
Eq. (1.27)	3	4	2	−1	1	0	0	5	$\frac{5}{3}$
Eq. (1.28)	4	2	1	0	0	−1	1	4	$\frac{4}{4}$
$-W''$	−7	−6	−3	1	0	1	0	−9	

Table 1.20

	y_1	y_2	y_3	y_4	y_5	y_6	y_7	b
Eq. (1.27)	0	$\frac{5}{2}$	$\frac{5}{4}$	-1	1	$\frac{3}{4}$	$-\frac{3}{4}$	2
Eq. (1.28)	1	$\frac{1}{2}$	$\frac{1}{4}$	0	0	$-\frac{1}{4}$	$\frac{1}{4}$	1
$-W''$	0	$-\frac{5}{2}$	$-\frac{5}{4}$	1	0	$-\frac{3}{4}$	$\frac{7}{4}$	-2

Table 1.21

	y_1	y_2	y_3	y_4	y_5	y_6	y_7	b
Eq. (1.27)	0	1	$\frac{1}{2}$	$-\frac{2}{5}$	$\frac{2}{5}$	$\frac{3}{10}$	$-\frac{3}{10}$	$\frac{4}{5}$
Eq. (1.28)	1	0	0	$\frac{1}{5}$	$-\frac{1}{5}$	$-\frac{2}{5}$	$\frac{2}{5}$	$\frac{4}{5}$
$-W''$	0	0	2	$\frac{2}{5}$	1	$\frac{16}{5}$	1	0

Phase II

Table 1.22

	y_1	y_2	y_3	y_4	y_6	b
Eq. (1.27)	0	1	$\frac{1}{2}$	$-\frac{2}{5}$	$\frac{3}{10}$	$\frac{4}{5}$
Eq. (1.28)	1	0	0	$\frac{1}{5}$	$-\frac{2}{5}$	$\frac{3}{5}$
$-W''$	14	8	6	0	0	0

Table 1.23

	y_1	y_2	y_3	y_4	y_6	b
Eq. (1.27)	0	1	$\frac{1}{2}$	$-\frac{2}{5}$	$\frac{3}{10}$	$\frac{4}{5}$
Eq. (1.28)	1	0	0	$\frac{1}{5}$	$-\frac{2}{5}$	$\frac{3}{5}$
$-W''$	0	0	2	$\frac{2}{5}$	$\frac{16}{5}$	$\frac{74}{5}$

Once the table at the beginning of phase II is put into canonical form it is apparent that no further iterations are necessary. The optimal solution is, ignoring artificial variables,

$$y_1^* = \tfrac{3}{5}, \quad y_2^* = \tfrac{4}{5}, \quad y_3^* = 0, \quad y_4^* = 0, \quad y_6^* = 0, \quad \text{and} \quad W''^* = \tfrac{74}{5}.$$

It is instructive to solve P_1 by the simplex method and compare the final table (Table 1.24 with Table 1.23).

Table 1.24

	x_1	x_2	x_3	x_4	x_5	b
Eq. (1.23)	0	1	$\frac{2}{5}$	$-\frac{3}{10}$	0	$\frac{16}{5}$
Eq. (1.24)	1	0	$-\frac{1}{5}$	$\frac{1}{10}$	0	$\frac{2}{5}$
Eq. (1.25)	0	0	0	$-\frac{1}{2}$	1	2
$-Z$	0	0	$\frac{3}{5}$	$\frac{4}{5}$	0	$\frac{74}{5}$

As the dual has the same data as the primal (but in rearranged form) it comes as no surprise that there are the following similarities between the two tables:

(1) $W^* = Z^*$; that is, the value of the optimal solution of the primal and of the dual are equal.
(2) The slack variable in constraint j in one problem has the value appearing in the bottom row, in the jth position, of the other problem.

Complementary Slackness

(3a) If a structural variable is positive, the slack variable in the corresponding constraint in the other problem is zero.
(3b) If a slack variable is positive, the structural variable (in the other problem) corresponding to its constraint is zero.

The relationships hold for any pair of primal–dual LP problems which have finite solution values.

Let us now illustrate these properties for the primal–dual example as shown in Tables 1.23 and (1.24):

(1) $W^* = Z^* = \frac{74}{5}$.
(2) Consider the slack variable x_3 in P_1. Its constraint (1.23) is associated with y_1 in the construction of D_1. Looking at Table 1.24 we can see that

$$x_3^* = 0,$$

which equals the $-W$ entry in the y_1 column.

Similarly,

$$x_4^* = 0,$$

which is associated with y_2, which equals the $-W$ entry in the y_2 column. x_5 is associated with y_3 and

$$x_5^* = 2,$$

which is the bottom-row entry in the y_3 column. The relationship also holds when we examine the dual slack variables

$$y_4^* = 0.$$

Its constraint is (1.27). If we considered D_1 as the primal and constructed P_1 as its dual, x_1 would be associated with (1.27). We can see in Table 1.24 that the $-Z$ row entry in the x_1 column is also zero.
Similarly,

$$y_6^* = 0,$$

and the $-Z$ row entry in the x_2 column is also zero.

(3a) x_1 is a structural variable with positive value ($\frac{2}{5}$). It is associated with constraint (1.27) in the dual. The slack variable in (1.27) is y_4, which is zero.
Similarly,

$$x_2^* > 0 \Rightarrow y^*{}_6 = 0.$$

Also, y_1^* is a structural variable with positive value ($\frac{3}{5}$). It is associated with (1.23) in the primal. The slack variable in that constraint is x_3 and $x_3^* = 0$.
Similarly,

$$y_2^* = \tfrac{4}{5} > 0$$

and

$$x_4^* = 0.$$

(3b) $x_5^* > 0$. Now x_5, from constraint (1.25), is associated with y_3 and $y_3^* = 0$.

We now state some useful duality theorems.

Theorem 1.1. *If P and D have corresponding feasible solutions X and Y, respectively, then*

$$\mathbf{CX} \leq \mathbf{b}^T\mathbf{Y}. \tag{1.29'}$$

This result implies that if a primal LP problem has a maximization objective then the value of any of its feasible solutions is a lower bound on the value of any corresponding feasible solution to the dual. Obviously, the value of any feasible solution to the dual is an upper bound on the value of any corresponding feasible solution to the primal.

Theorem 1.2. *Equality holds in* (1.29′) *if and only if* $\mathbf{X}$ *and* $\mathbf{Y}$ *are optimal solutions for P and D, respectively.*

We now use these results to develop a new method for solving LP problems. It is called the *dual simplex method*.

1.1.7. The Dual Simplex Method

1.1.7.1 *Background*. Consider an LP problem P, which has a finite optimal solution and is solved by the simplex method. The final table can be interpreted to find the optimal solution to the dual D, of P. Indeed every table in canonical form produced in the course of phase II of the simplex method can be interpreted to find a solution to the dual. Thus as the simplex method is producing a sequence of solutions for P it is also producing a sequence of solutions for D. It turns out that all but the last in this sequence will be infeasible for D. The values of the solutions are progressively worse (decreasing if D maximizes or increasing if D minimizes). The final solution is optimal and by definition feasible.

This can be illustrated by solving P_1 of Section 1.1.6 by the simplex method. On the introduction of slack variables x_1, x_2, and x_3, P_1 becomes:

$$\text{Maximize } 5x_1 + 4x_2 \qquad (=Z)$$

subject to

$$3x_1 + 4x_2 + x_3 \qquad = 14, \tag{1.23'}$$

$$4x_1 + 2x_2 \qquad + x_4 \qquad = 8, \tag{1.24'}$$

$$2x_1 + x_2 \qquad + x_5 = 6, \tag{1.25'}$$

$$x_1, x_2 \geq 0.$$

Tables necessary to produce the optimal solution are Tables 1.25–1.27.

Table 1.25

	x_1	x_2	x_3	x_4	x_5	b	Ratios
Eq. (1.23′)	3	4	1	0	0	14	$\frac{14}{3}$
Eq. (1.24′)	4	2	0	1	0	8	$\frac{8}{4}$
Eq. (1.25′)	2	1	0	0	1	6	$\frac{6}{2}$
$-Z$	-5	-4	0	0	0	0	

Table 1.26

	x_1	x_2	x_3	x_4	x_5	b	Ratios
Eq. (1.23′)	0	$\frac{5}{2}$	1	$-\frac{3}{4}$	0	8	$\frac{16}{5}$
Eq. (1.24′)	1	$\frac{1}{2}$	0	$\frac{1}{4}$	0	2	$\frac{4}{1}$
Eq. (1.25′)	0	0	0	$-\frac{1}{2}$	1	2	
$-Z$	0	$-\frac{3}{2}$	0	$\frac{5}{4}$	0	10	

Table 1.27

	x_1	x_2	x_3	x_4	x_5	b
Eq. (1.23′)	0	1	$\frac{2}{5}$	$-\frac{3}{10}$	0	$\frac{16}{5}$
Eq. (1.24′)	1	0	$-\frac{1}{5}$	$\frac{1}{10}$	0	$\frac{2}{5}$
Eq. (1.25′)	0	0	0	$-\frac{1}{2}$	1	2
$-Z$	0	0	$\frac{3}{5}$	$\frac{4}{5}$	0	$\frac{74}{5}$

The dual of P_1 is D_1:

$$\text{Minimize } 14y_1 + 8y_2 + 6y_3 \quad (=W)$$

subject to

$$\begin{aligned} 3y_1 + 4y_2 + 2y_3 - y_4 &= 5 \\ 4y_1 + 2y_2 + y_3 - y_6 &= 4, \\ y_1, y_2, y_3, y_4, y_6 &\geq 0. \end{aligned}$$

We can interpret Tables (1.25)–(1.27) using the complementary slackness rules (3a) and (3b) stated in Section 1.1.6 to find solutions to P_1:

$y_1 = 0,\ y_2 = 0,\ y_3 = 0,\ y_4 = -5,\ y_6 = -4,$ and $W = 0$ (from Table 1.25);

$y_1 = 0,\ y_2 = \frac{5}{4},\ y_3 = 0,\ y_4 = 0,\ y_6 = -\frac{3}{2},$ and $W = 10$ (from Table 1.26);

$y_1 = \frac{3}{5},\ y_2 = \frac{4}{5},\ y_3 = 0,\ y_4 = 0,\ y_6 = 0,$ and $W = \frac{74}{5}$ (from Table 1.27).

The first two of these solutions are infeasible as they possess negative values. The solutions get progressively worse in value; that is, W increases for a minimization problem.

The *dual simplex method* is a method for solving LP problems, which is based on the above observations. It is quite different from merely taking the dual of an LP problem, solving it by the simplex method, and then interpreting the final table. It solves an LP problem P by performing the same calculations as the simplex method would produce when used to solve the dual of P. This means that the dual simplex method produces a sequence of infeasible solutions with progressively worsening values and then a final, optimal solution. The method is also used as a subroutine to regain optimality when an additional constraint renders an optimal solution no longer feasible. This is especially useful in the cutting-plane methods for integer programming, as described in Section 2.1. We now illustrate the dual simplex method with a numerical example.

1.1.7.2. *A Numerical Example.* We shall now solve D_1, discussed in the last section, by the dual simplex method.

$$\text{Maximize } -14y_1 - 8y_2 - 6y_3 \quad (= W'')$$

subject to

$$3y_1 + 4y_2 + 2y_3 - y_4 = 5 \qquad (1.26')$$

$$4y_1 + 2y_2 + y_3 - y_6 = 4, \qquad (1.27')$$

$$y_1, \ldots, y_4, y_6 \geq 0.$$

Note that we assume a maximization objective. The initial table is shown in Table 1.28. The method does not introduce artificial variables, so we must find another way of finding an initial basic solution. This is achieved by multiplying the rows in the table other than $-W''$ by -1. This produces Table 1.29.

Table 1.28

	y_1	y_2	y_3	y_4	y_6	b
Eq. (1.26′)	3	4	2	−1	0	5
Eq. (1.27′)	4	2	1	0	−1	4
$-W''$	14	8	6	0	0	0

Table 1.29

	y_1	y_2	y_3	y_4	y_6	b
Eq. (1.26′)	−3	(−4)	−2	1	0	−5
Eq. (1.27′)	−4	−2	−1	0	1	−4
$-W''$	14	8	6	0	0	0

We now have an initial basic solution:

$$y_4 = -5,$$
$$y_6 = -4,$$
$$y_1 = y_2 = y_3 = 0,$$
$$W'' = 0.$$

This solution also satisfies the criterion for optimality, namely, that all the $-W''$ row entries are nonnegative. Each solution produced by the dual simplex method will satisfy this criterion. Of course this solution is not optimal as it is not feasible—it contains negative values. It corresponds to the initial table for P_1 when it is solved by the simplex method—Table 1.25. Its value $W'' = 0$ is better than the known optimal value of $\frac{74}{5}$. (Remember D_1 was originally a minimization problem.)

In the simplex method, we first decide which variable enters the basis and then decide which variable leaves. In the dual simplex method, we first decide which variable leaves the basis. This is settled by removing the variable which is the most negative, in this case $y_4 = -5$. If all variables are nonnegative, the current solution is optimal as it satisfies the criterion for optimality and the method is terminated. The identification of the b entry with the negative entry large in magnitude has a sequel in the simplex method. It corresponds to scanning the $-Z$ row for the most negative entry in order to settle which variable enters the basis. For P in Table 1.25, it means choosing the same number, -5, to bring x_1 into the basis. Note that y_4, the departing variable for D_1, is the slack variable of constraint (1.26) which is associated with x_1.

We now settle which variable replaces y_4 in the basis. To do this we take the same ratios as those formed in the simplex method. Those ratios were

$$(b_1/a_{11}, b_2/a_{12}, b_3/a_{13}) = (14/3, 8/4, 6/2).$$

Only ratios with positive denominators were formed. The corresponding ratios in Table 1.29 are

$$\tfrac{14}{-3},\ \tfrac{8}{-4},\ \text{and}\ \tfrac{6}{-2}.$$

Only ratios with negative denominators are formed. The same ratios as selected in the simplex method are selected here. Because the denominators are negative this means that the largest (least negative) ratio is selected. So that $\frac{8}{-4}$ is selected and y_2 is to enter the basis. The -4 entry is circled in Table 1.29 as the pivot and a change of basis is performed using Gauss–Jordon elimination in the usual way. This produces Table 1.30.

Table 1.30

	y_1	y_2	y_3	y_4	y_6	b
Eq. (1.26′)	$\frac{3}{4}$	1	$-\frac{1}{2}$	$-\frac{1}{4}$	0	$\frac{5}{4}$
Eq. (1.27′)	$\textcircled{-\frac{5}{2}}$	0	0	$-\frac{1}{2}$	1	$-\frac{3}{2}$
$-W''$	8	0	2	2	0	-10

Please compare this table with Table 1.26. It contains a new solution for D_1:

$$y_2 = \tfrac{5}{4},$$
$$y_6 = -\tfrac{3}{2},$$
$$y_1 = y_3 = y_4 = 0,$$
$$W'' = 10.$$

This solution is still infeasible as y_6 is negative and it is worse in value than the last solution. However, y_2 is nonnegative so we are closer to feasibility. (We are also closer to optimality as the criterion for optimality is still satisfied.)

The same steps are now repeated on Table 1.30. Because y_6 is the only negative variable, it is fated to leave the basis. The ratios $8/(-5/2)$ and $2/(-1/2)$ are compared to find the incoming variable. The third ratio is not taken because it has a zero (and hence nonnegative) denominator. The variable y_1 comes into the basis, producing Table 1.31.

Table 1.31

	y_1	y_2	y_3	y_4	y_6	b
Eq. (1.26′)	0	1	0	$-\frac{7}{20}$	$-\frac{3}{10}$	$\frac{4}{5}$
Eq. (1.27′)	1	0	0	$\frac{1}{5}$	$-\frac{2}{5}$	$\frac{3}{5}$
$-W''$	0	0	2	$\frac{2}{5}$	$\frac{16}{5}$	$-\frac{74}{5}$

The reader should compare this table with Table 1.27. It contains the optimal solution for D_1:

$$y_1 = \tfrac{3}{5},$$

$$y_2 = \tfrac{4}{5},$$

$$y_3 = y_4 = y_6 = 0,$$

$$W = \tfrac{74}{5}.$$

It is optimal because it is basic and feasible; all the variables are nonnegative, and it satisfies the criterion for optimality.

1.1.7.3. *A Summary of the Dual Simplex Method.* Here are the steps of the dual simplex method when applied to a problem in standard form.

Steps

(1) Multiply each constraint containing a slack variable with negative coefficient by -1. The initial basis is the set of slack variables.
(2) If the basic solution has all nonnegative values, it is optimal. In this case terminate the procedure. Otherwise proceed to step (3).
(3) Set up a table in canonical form for the initial basic solution. Identify all variables with negative values. Among these select the one whose value is largest in magnitude to leave the basis.
(4) Identify the row in the table which has a $+1$ coefficient for the departing variable. Identify all negative entries in this row. For each such entry, form a ratio of the entry at the bottom of its column over

itself. The column corresponding to the largest (least negative) of these ratios corresponds to the variable entering the basis.

(5) Make a change of basis using the Gauss–Jordon elimination. Go back to step (2).

1.1.7.4. *One More Numerical Example for the Dual Simplex Method*. The following problem will now be solved using the dual simplex method:

$$\text{Maximize } -6y_1 - 8y_2 - 9y_3 - 12y_4 \quad (=W)$$

subject to

$$y_1 + 3y_2 + 2y_3 + 2y_4 - y_5 = 3, \tag{1.32}$$

$$2y_1 + 4y_2 + 3y_3 + y_4 - y_6 = 2, \tag{1.33}$$

$$y_1 + 2y_2 + 3y_3 + 2y_4 - y_7 = 1, \tag{1.34}$$

$$3y_1 + y_2 + y_3 + 2y_4 - y_8 = 2, \tag{1.35}$$

$$y_1, y_2, \ldots, y_8 \geq 0.$$

The reader is urged to formulate the dual, solve it by the simplex method, and compare the sequence of tables produced with Tables 1.32–1.34.

Table 1.32

	y_1	y_2	y_3	y_4	y_5	y_6	y_7	y_8	b
Eq. (1.32)	-1	(-3)	-2	-2	1	0	0	0	-3
Eq. (1.33)	-2	-4	-3	-1	0	1	0	0	-2
Eq. (1.34)	-1	-2	-3	-2	0	0	1	0	-1
Eq. (1.35)	-3	-1	-1	-2	0	0	0	1	-2
$-W$	6	8	9	12	0	0	0	0	0
Ratios	$-\frac{6}{1}$	$-\frac{8}{3}$	$-\frac{9}{2}$	$-\frac{12}{2}$					

Table 1.33

	y_1	y_2	y_3	y_4	y_5	y_6	y_7	y_8	b
Eq. (1.32)	$\frac{1}{3}$	1	$\frac{2}{3}$	$\frac{2}{3}$	$-\frac{1}{3}$	0	0	0	1
Eq. (1.33)	$-\frac{2}{3}$	0	$-\frac{1}{3}$	$\frac{5}{3}$	$-\frac{4}{3}$	1	0	0	2
Eq. (1.34)	$-\frac{1}{3}$	0	$-\frac{5}{3}$	$-\frac{2}{3}$	$-\frac{2}{3}$	0	1	0	1
Eq. (1.35)	$(-\frac{8}{3})$	0	$-\frac{1}{3}$	$-\frac{4}{3}$	$-\frac{1}{3}$	0	0	1	-1
$-W$	$\frac{10}{3}$	0	$\frac{11}{3}$	$\frac{20}{3}$	$\frac{8}{3}$	0	0	0	-8
Ratios	$-\frac{5}{4}$		$-\frac{11}{1}$	$-\frac{5}{1}$	$-\frac{8}{1}$				

Table 1.34

	y_1	y_2	y_3	y_4	y_5	y_6	y_7	y_8	b
Eq. (1.32)	0	1	$\frac{5}{8}$	$\frac{1}{2}$	$-\frac{3}{8}$	0	0	$\frac{1}{8}$	$\frac{7}{8}$
Eq. (1.33)	0	0	$-\frac{1}{4}$	2	$-\frac{5}{4}$	1	0	$-\frac{1}{4}$	$\frac{9}{4}$
Eq. (1.34)	0	0	$-\frac{13}{8}$	$-\frac{1}{2}$	$-\frac{5}{8}$	0	1	$-\frac{1}{8}$	$\frac{9}{8}$
Eq. (1.35)	1	0	$\frac{1}{8}$	$\frac{1}{2}$	$\frac{1}{8}$	0	0	$-\frac{3}{8}$	$\frac{3}{8}$
$-W$	0	0	$\frac{13}{4}$	5	$\frac{9}{4}$	0	0	$\frac{5}{4}$	$-\frac{37}{4}$

The optimal solution is

$$y_1^* = \tfrac{3}{8}, \quad y_2^* = \tfrac{7}{8}, \quad y_6^* = \tfrac{9}{4}, \quad \text{and} \quad y_7^* = \tfrac{9}{8};$$

$$y_3^* = y_4^* = y_5^* = y_8^* = 0;$$

$$W^* = -\tfrac{37}{4}.$$

1.2. The Transportation Problem

1.2.1. A Simple Numerical Example

The Quickly But Surely shipping company has a problem. It has to supply bags of cement to four warehouses from the three factories of one of its clients. The daily demands of the four warehouses w_1, w_2, w_3, and w_4 are 30, 20, 35, and 20, respectively. The daily output of the three factories; f_1, f_2, and f_3, are 40, 40, and 25, respectively. The cost of shipping one bag from each factory to each warehouse is given below:

		Warehouses			
		w_1	w_2	w_3	w_4
Factories	f_1	6	5	7	9
	f_2	3	2	4	1
	f_3	7	3	9	5

The shipping manager of Quickly But Surely has to devise a schedule that ensures the following: each warehouse receives at least as much as its daily demand; each factory does not have to ship more than its daily output; and the total shipping cost is as small as possible. To this end, we define some decision variables. Let

x_{ij} = the number of bags shipped daily from factory i to warehouse j,

$$i=1,2,3 \quad \text{and} \quad j=1,2,3,4.$$

What would be the cost of a policy defined by giving values to the x_{ij} variables? We begin with x_{11}, the number shipped from f_1 to w_1. Shipping one bag costs six units. As the costs are linear we can say that shipping x_{11} bags will cost $6x_{11}$ units. By creating similar expressions for each factory warehouse pair we come to the fact that the total shipping cost is

$$6x_{11}+5x_{12}+7x_{13}+9x_{14}+3x_{21}+2x_{22}$$
$$+4x_{23}+x_{24}+7x_{31}+3x_{32}+9x_{33}+5x_{34}.$$

It may be that a certain factory, say f_i, will not supply a certain warehouse, say w_j, in the minimum cost schedule. In this case, $x_{ij}=0$ and the unit shipping cost from f_i to w_j will not contribute to the total bill. The manager must minimize the value of this expression.

We now describe some constraints on this minimization process. How much is f_1 asked to ship? It ships x_{11} bags to w_1, x_{12} to w_2, x_{13} to w_3, and x_{14} to w_4; or

$$x_{11}+x_{12}+x_{13}+x_{14}$$

in all. We know that this sum cannot exceed the daily output of f_1, namely, 40 bags. Thus we have our first constraint:

$$x_{11}+x_{12}+x_{13}+x_{14}\le 40.$$

There are similar constraints on the output of f_2 and f_3, namely,

$$x_{21}+x_{22}+x_{23}+x_{24}\le 40$$

and

$$x_{31}+x_{32}+x_{33}+x_{34}\le 25.$$

There is a second set of constraints concerning the demands of the warehouses. Let us begin with w_1. What is its total intake? It receives x_{11} bags from f_1, x_{21} from f_2, and x_{31} from f_3, or

$$x_{11}+x_{21}+x_{31}$$

in all. We know that this sum must be at least the daily demand of w_1, namely, 30 bags. Thus we have the constraint:

$$x_{11}+x_{21}+x_{31}\ge 30.$$

There are similar constraints on the intake of w_2, w_3, and w_4, namely,

$$x_{12}+x_{22}+x_{32}\ge 20,$$
$$x_{13}+x_{23}+x_{33}\ge 35,$$
$$x_{14}+x_{24}+x_{34}\ge 20.$$

Naturally, the amount shipped from any factory f_i to any warehouse w_j must be nonnegative:

$$x_{ij}\ge 0,\qquad i=1,2,3 \quad\text{and}\quad j=1,2,3,4.$$

We now have a complete mathematical model of the shipping manager's problem:

$$\text{Minimize } 6x_{11}+5x_{12}+7x_{13}+9x_{14}+3x_{21}+2x_{22} \\ +4x_{23}+x_{24}+7x_{31}+3x_{32}+9x_{33}+5x_{34}$$

subject to

$$\begin{aligned} x_{11}+x_{12}+x_{13}+x_{14} &\leq 40, \\ x_{21}+x_{22}+x_{23}+x_{24} &\leq 40, \\ x_{31}+x_{32}+x_{33}+x_{34} &\leq 25, \\ x_{11}+x_{21}+x_{31} &\geq 20, \\ x_{12}+x_{22}+x_{32} &\geq 20, \\ x_{13}+x_{23}+x_{33} &\geq 35, \\ x_{ij}\geq 0, \quad i=1,2,3 &\text{ and } j=1,2,3,4. \end{aligned} \tag{1.36}$$

We can arrange the data of the problem as in Table 1.35. An appropriate unit cost appears in a little box in the top left-hand corner of each cell; the warehouse demands are listed along the bottom; and the factory outputs in the rightmost column.

Table 1.35

		Warehouses				
		w_1	w_2	w_3	w_4	Outputs
	f_1	6	5	7	9	40
Factories	f_2	3	2	4	1	40
	f_3	7	3	9	5	25
Demands		30	20	35	20	

Does a feasible solution exist? That is, can we find values for all the x_{ij}'s which satisfy all the constraints? We can begin to answer this question by adding up the total demand of all warehouses and finding out whether or not it is less than the total output of the factories. If it is not, we shall be unable to find a feasible solution. As it turns out the total demand is $30+20+35+20=105$ bags. The total output is also 105 bags so that it is possible to solve the problem satisfactorily. The methods that we shall use to solve problems of this type assume that total demand T_d, equals total output T_o, as it does in our example. What is to be done to convert an *unbalanced* problem (where $T_d \neq T_o$) into a *balanced* one so that the methods can be applied? We shall answer this via our example problem. Let us assume temporarily that the demand of w_4 is only 10 rather than 20.

Then $T_d = 95 < 105 = T_o$ and we have an unbalanced problem. In this situation, we introduce a "dummy" (fictitious) warehouse w_5 with total demand equal to $T_o - T_d = 10$. The unit transportation costs from each factory to w_5 are defined to be zero. We can expand the original table (Table 1.35) to include w_5 (see Table 1.36):

Table 1.36

		Warehouses					
		w_1	w_2	w_3	w_4	w_5	Outputs
Factories	f_1	6	5	7	9	0	40
	f_2	3	2	4	1	0	40
	f_3	7	3	9	5	0	25
Demands		30	20	35	10	10	

We now assume that w_4 has its original demand of 20 and temporarily assume that f_3 has an output reduced from 40 to 20. This causes an imbalance of $T_d - T_o = 20$. Obviously, the problem now does not have a feasible solution as total output cannot match total demand. However, there are problems which have the same structure as Eqs. (1.36) but have nothing to do with factories and warehouses. Many of these still have feasible solutions when $T_d > T_o$. We shall see an example of this in Section 1.3. In these cases we introduce a dummy "factory" with output equal to $T_d - T_o$. In the present instance this means introducing f_4 with output 20. The unit transportation costs from it to each warehouse are defined to be zero. We can expand the original table (Table 1.35) to include f_4 (see Table 1.37):

Table 1.37

		Warehouses				
		w_1	w_2	w_3	w_4	Outputs
Factories	f_1	6	5	7	9	40
	f_2	3	2	4	1	40
	f_3	7	3	9	5	25
	f_4	0	0	0	0	20
Demands		30	20	35	20	

Any solution to a problem by the introduction of a dummy can be converted into a solution for the original problem. This is done simply by ignoring the shipments to or from the dummy. The total cost of the solutions with and without the dummy are the same because the shipment costs to or from the dummy are zero.

1.2.2. The General Formulation

The problem just discussed is an example of what is called the *transportation problem*. In its general form this problem has the following parameters:

$$\begin{aligned} m &= \text{the number of factories;} \\ n &= \text{the number of warehouses;} \\ a_i &= \text{the daily output of } f_i, \qquad i=1,2,\ldots,m; \\ b_j &= \text{the daily demand of } w_j, \qquad j=1,2,\ldots,n; \\ c_{ij} &= \text{the unit transportation cost from } f_i \text{ to } w_j. \end{aligned}$$

The problem is to

$$\text{Minimize } \sum_{i=1}^{m} \sum_{j=1}^{n} c_{ij} x_{ij}$$

subject to

$$\begin{aligned} \sum_{i=1}^{m} x_{ij} &\geq b_j, \qquad j=1,2,\ldots,n; \\ \sum_{j=1}^{n} x_{ij} &\leq a_i, \qquad i=1,2,\ldots,m; \\ x_{ij} &\geq 0, \qquad i=1,2,\ldots,m, \quad j=1,2,\ldots,n. \end{aligned} \tag{1.37}$$

Table 1.38 represents this general problem:

Table 1.38

		Warehouses				
		w_1	w_2	$\cdots$	w_n	Outputs
	f_1	c_{11}	c_{12}	$\cdots$	c_{1n}	a_1
Factories	f_2	c_{21}	c_{22}	$\cdots$	c_{2n}	a_2
	$\vdots$	$\vdots$	$\vdots$		$\vdots$	$\vdots$
	f_m	c_{m1}	c_{m2}	$\cdots$	c_{mn}	a_m
Demands		b_1	b_2	$\cdots$	b_n	

This problem is a linear-programming problem and as such can be solved by the simplex method (see Section 1.1). However, it can be solved more efficiently by a modification of the simplex method designed to take advantage of the problem's special structure. The structure is:

(1) The coefficient of each decision variable in each constraint is either zero or one.
(2) The constant on the right-hand side of each constraint is assumed to be an integer.
(3) The matrix **A** of the constraint set (see Section 1.1 for an explanation of this term) has a distinctive pattern.

Structure feature (3) is best explained via our numerical example. If we rearrange the constraints in (1.36) so that each variable has its own column we can express the set of left-hand sides of the constraints in the form:

$$\begin{pmatrix} 1 & 1 & 1 & 1 & 0 & 0 & 0 & 0 & 0 & 0 & 0 & 0 \\ 0 & 0 & 0 & 0 & 1 & 1 & 1 & 1 & 0 & 0 & 0 & 0 \\ 0 & 0 & 0 & 0 & 0 & 0 & 0 & 0 & 1 & 1 & 1 & 1 \\ 1 & 0 & 0 & 0 & 1 & 0 & 0 & 0 & 1 & 0 & 0 & 0 \\ 0 & 1 & 0 & 0 & 0 & 1 & 0 & 0 & 0 & 1 & 0 & 0 \\ 0 & 0 & 1 & 0 & 0 & 0 & 1 & 0 & 0 & 0 & 1 & 0 \\ 0 & 0 & 0 & 1 & 0 & 0 & 0 & 1 & 0 & 0 & 0 & 1 \end{pmatrix} \begin{pmatrix} x_{11} \\ x_{12} \\ x_{13} \\ x_{14} \\ x_{21} \\ x_{22} \\ x_{23} \\ x_{24} \\ x_{31} \\ x_{32} \\ x_{33} \\ x_{34} \end{pmatrix} = \mathbf{AX},$$

where **A** is the 7×12 matrix and **X** is the column vector of decision variables. Let us examine **A**. The first three rows, one row for each factory, each has a block of 1's—a 1 for each warehouse. The remaining four rows, one row for each warehouse, each has a 1 for each factory. This basic pattern will remain the same if we increase the number of factories or warehouses or both. As an exercise the reader should increase m and n each by 1 in the example problem and discover the new pattern of **A** and then find the general description of **A** for Eqs. (1.37).

An LP problem with structure features (1), (2), and (3) has a very important property. If the problem has a feasible solution there will always exist an optimal solution with integer-valued decision variables. This property is the cornerstone of the efficiency of the modified simplex method. The method needs as input an initial feasible solution, which it then transforms to optimality. So we begin in the next section with methods which produce an initial feasible solution.

1.2.3. Finding an Initial Feasible Solution

We represent any solution to the transportation problem by inserting the value of x_{ij} in the cell in the ith row and the jth column—usually with a circle around it. Thus we wish to find a set of numbers (representing the x_{ij}'s) for the cells of the table so that the sum of these for each column j equals b_j and the sum for each row i equals a_i. This equality arises because in a balanced problem each factory will ship all of its output and each warehouse will receive exactly its demand. The reader should follow the explanations that follow by making a copy of Table 1.35 and putting in the numbers as they are calculated. The methods for finding a feasible solution will be explained via our example problem.

1.2.3.1. *The Northwest-Corner Method.* We begin by allocating as many bags as possible to the cell in the "northwest" corner of Table 1.35. This is the cell in the top left-hand corner. The reader should refer to Table 1.39, ignoring the uncircled numbers. In our case the maximum that can be allocated is the minimum of the output of f_1, that is, $a_1 = 40$ and the demand of w_1, that is, $b_1 = 30$. So we allocate 30 and w_1 is satisfied. We can remove the first column from consideration as the number in it equals b_1. The output of f_1 is now reduced from 40 to 10. This creates a new northwest corner—the x_{12} cell. We can allocate 10 to it, the minimum of the revised output of f_1 and the demand of w_2. The f_1 row is now removed as its total output is allocated. The demand of w_2 is reduced to 10. The new northwest corner is the cell of x_{22}. We allocate 10 to it and can remove the second column. Proceeding in this way we eventually produce the schedule shown in Table 1.39. It corresponds to the following solution:

$$x_{11} = 30, \quad x_{12} = 10, \quad x_{22} = 10, \quad x_{23} = 30,$$
$$x_{33} = 5, \quad \text{and} \quad x_{34} = 20.$$

All other variables are zero.

Northwest Corner

Table 1.39

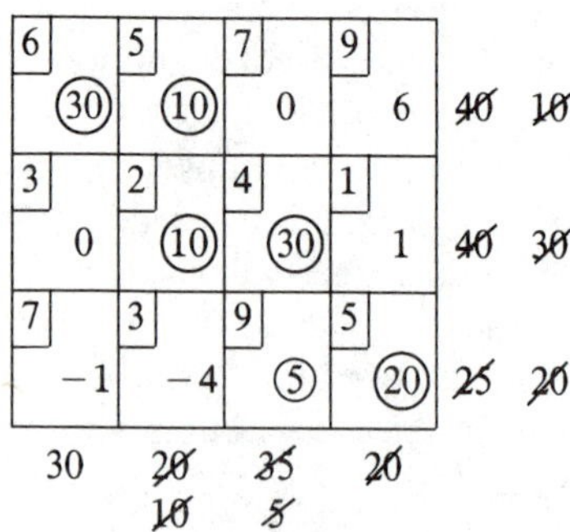

This implies that f_1 should ship 30 units to w_1 and 10 to w_2; f_2 should ship 10 to w_2 and 30 to w_3; and f_3 should ship 5 to w_3 and 20 to w_4. The total cost of the schedule is the sum of the products of the numbers circled times the unit costs in their cells:

$$(6\times 30)+(5\times 10)+(2\times 10)+(4\times 30)+(9\times 5)+(5\times 20)=515 \text{ units.}$$

We shall refer to Tables 1.40 and 1.41 later.

Table 1.40

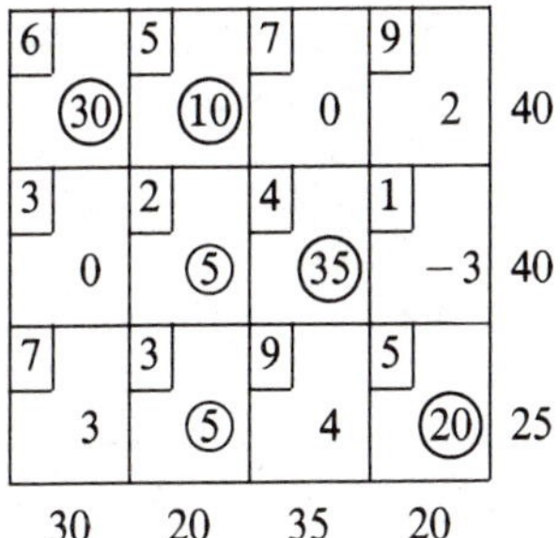

[6] (30)	[5] (10)	[7] 0	[9] 2	40
[3] 0	[2] (5)	[4] (35)	[1] −3	40
[7] 3	[3] (5)	[9] 4	[5] (20)	25
30	20	35	20	

Table 1.41(a)

[6] (30)	[5] (10)	[7] −3	[9] 2	40
[3] 3	[2] 3	[4] (35)	[1] (5)	40
[7] 3	[3] (10)	[9] 1	[5] (15)	25
30	20	35	20	

Table 1.41(b)

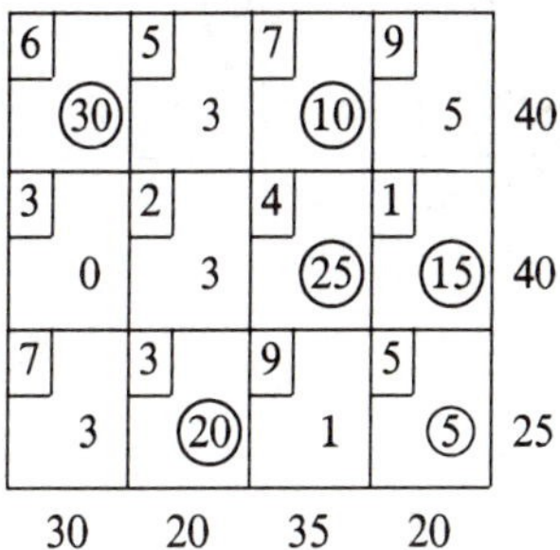

[6] (30)	[5] 3	[7] (10)	[9] 5	40
[3] 0	[2] 3	[4] (25)	[1] (15)	40
[7] 3	[3] (20)	[9] 1	[5] (5)	25
30	20	35	20	

Alternative Optimum

Table 1.41(c)

[6] (5)	[5]	[7] (35)	[9]	40
[3] (25)	[2]	[4]	[1] (15)	40
[7]	[3] (20)	[9]	[5] (5)	25
30	20	35	20	

1.2.3.2. *The Least-Cost Method.* The reader should refer to Tables 1.42(a) and 1.42(b), ignoring the uncircled numbers and the circled zero. We begin by identifying the smallest unit cost in the table and allocating as much as we can to its cell. In our case the least cost is $c_{24}=1$. We allocate 20 to its cell as it is the minimum of $a_2=40$ and $b_4=20$. The fourth column is removed from consideration as it is satisfied and a_2 is reduced to 20. The next smallest cost is c_{22}. We allocate the most we can to its cell, which is 20. This is because both the reduced output of f_2 and the demand of w_2 are 20. Row 2 and column 2 are now both removed. Continuing in this way we eventually produce Table 1.42(a). This solution has cost:

$$(6\times 30)+(2\times 20)+(7\times 10)+(9\times 25)+(1\times 20)=535 \text{ units.}$$

Table 1.42(a)

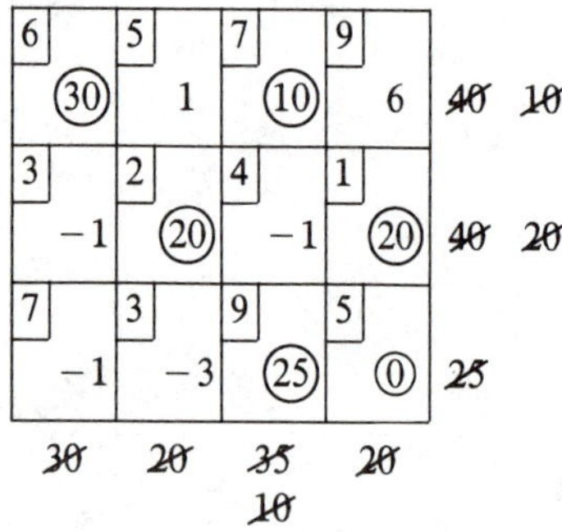

[6] (30)	[5] 1	[7] (10)	[9] 6	~~40~~ ~~10~~
[3] −1	[2] (20)	[4] −1	[1] (20)	~~40~~ ~~20~~
[7] −1	[3] −3	[9] (25)	[5] (0)	~~25~~
~~30~~	~~20~~	~~35~~ ~~10~~	~~20~~	

Table 1.42(b)

[6] (30)	[5] 4	[7] (10)	[9] 9	40
[3] 2	[2] (20)	[4] −4	[1] (20)	40
[7] −1	[3] (0)	[9] (25)	[5] 3	25
30	20	35	20	

Least Cost

We shall refer to Tables 1.43 later.

Table 1.43(a)

[6] (30)	[5] 4	[7] (10)	[9] 5	40
[3] 0	[2] 4	[4] (20)	[1] (20)	40
[7] −1	[3] (20)	[9] (5)	[5] −1	25
30	20	35	20	

Table 1.43(b)

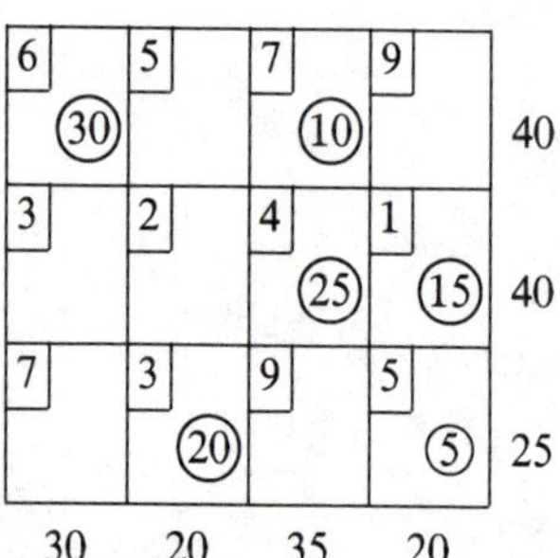

[6] (30)	[5]	[7] (10)	[9]	40
[3]	[2]	[4] (25)	[1] (15)	40
[7]	[3] (20)	[9]	[5] (5)	25
30	20	35	20	

This solution is worse in value than that produced by the northwest corner method. We now introduce a method which usually produces optimal or near-optimal initial feasible solutions, but at the price of more computation.

1.2.3.3. *The Vogel-Approximation Method.* The reader should refer to Tables 1.44–1.49. We begin by calculating column penalties. This is achieved by identifying the smallest and second smallest cost in each column. The difference between the two is defined to be the penalty for the column. As an example, look at the first column. The smallest cost is $c_{21} = 3$; the second smallest cost is $c_{11} = 6$. The penalty for column 1 is $c_{11} - c_{21} = 6 - 3 = 3$. The rationale for this is as follows. If we cannot allocate anything to the cell with the smallest cost in a column we must pay a penalty by having to allocate to an inferior cell. The column penalties are shown below the warehouse demands. Row penalties are calculated in the same way and are shown to the right of the rows. The allocation now begins. We find the cell with the largest sum of row and column penalties. As much as possible is allocated to it. In our case we allocate 20 to the x_{34} cell. This necessitates the removal of the w_4 column and the penalties must be calculated anew. The new penalties are shown in Table 1.45. The procedure continues in this way, as shown in Tables 1.44–1.49, where the complete solution is displayed. It has the value:

$$(3\times 30)+(5\times 15)+(3\times 5)+(7\times 25)+(4\times 10)+(5\times 20)=495 \text{ units.}$$

This represents an improvement over the last two solutions.

Vogel Approximation

Table 1.44

					Outputs	Row penalties
	6	5	7	9	40	1
	3	2	4	1	40	1
	7	3	9	5 (20)	25	2
Demands	30	20	35	20		
Column penalties	3	1	3	4		

New Penalties

Table 1.45

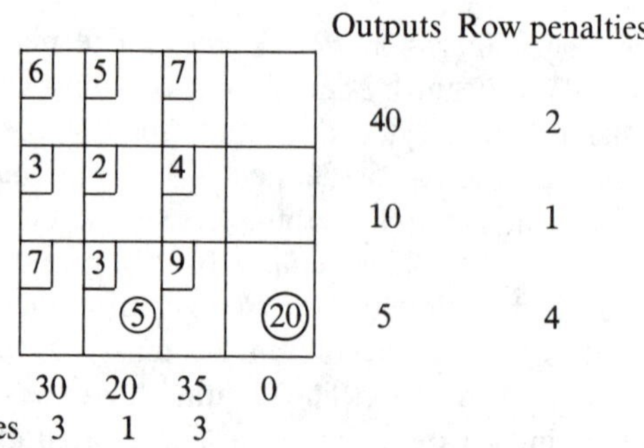

					Outputs	Row penalties
	6	5	7		40	2
	3	2	4		10	1
	7	3 (5)	9	(20)	5	4
Demands	30	20	35	0		
Column penalties	3	1	3			

Table 1.46

					Outputs	Row penalties
	6	5	7		40	2
	3 (30)	2	4		40	1
		(5)		(20)	0	
Demands	30	15	35	0		
Column penalties	3	3	3			

Table 1.47

					Outputs	Row penalties
		5	7		40	2
	(30)	2	4 (10)		10	2
		(5)		(20)	0	
Demands	0	15	35	0		
Column Penalties		3	3			

Table 1.48

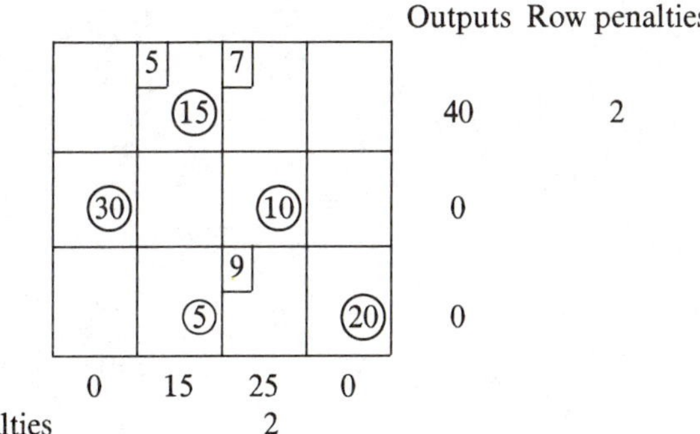

					Outputs	Row penalties
		[5] (15)	[7]		40	2
	(30)		(10)		0	
		(5)	[9]	(20)	0	
Demands	0	15	25	0		
Column Penalties			2			

The Final Allocation

Table 1.49

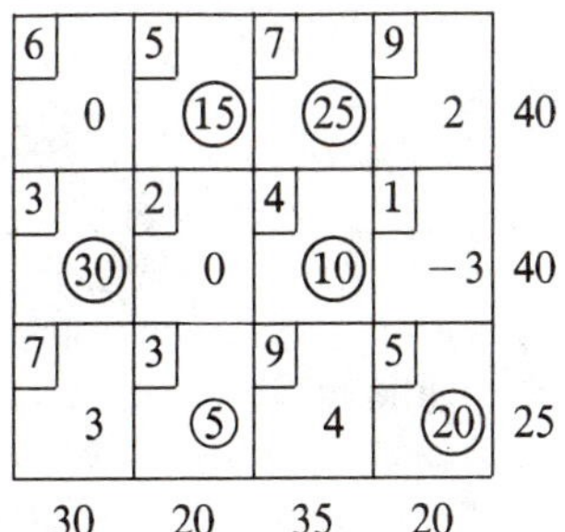

[6] 0	[5] (15)	[7] (25)	[9] 2	40
[3] (30)	[2] 0	[4] (10)	[1] −3	40
[7] 3	[3] (5)	[9] 4	[5] (20)	25
30	20	35	20	

We shall refer to Tables 1.50 and 1.51 later.

Table 1.50

[6] −1	[5] (5)	[7] (35)	[9] 2	40
[3] (30)	[2] 3	[4] 3	[1] (10)	40
[7] 0	[3] (15)	[9] 4	[5] (10)	25
30	20	35	20	

Table 1.51

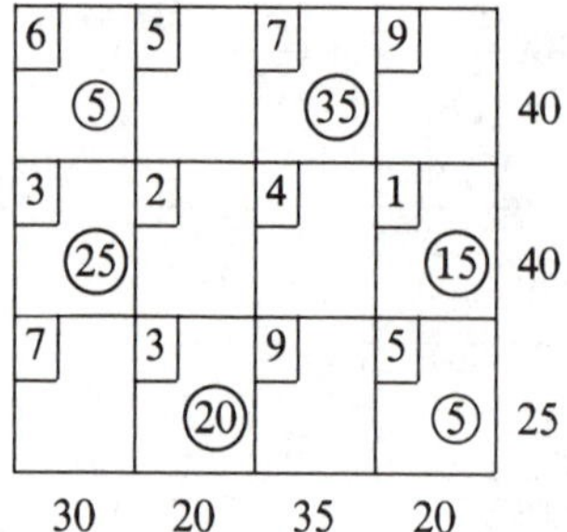

6	5	7	9	
(5)		(35)		40
3	2	4	1	
(25)			(15)	40
7	3	9	5	
	(20)		(5)	25
30	20	35	20	

1.2.4. Transportation-Problem Algorithm

Once we have a feasible solution for the transportation problem, we shall want to discover whether or not it is optimal. If it is not, we must then transform it to optimality. We achieve this by the *stepping-stone* method.

The Stepping-Stone Method

We explain the stepping-stone method by using it on Table 1.39. The first task is to establish whether or not the solution is optimal. We happen to know that Table 1.49 represents a cheaper solution so that Table 1.39 cannot be optimal. However, if the northwest-corner method alone has been used we must have a way of settling optimality. We begin by asking the following question for each empty cell. If exactly one unit was assigned to *only* that cell, would the total shipping cost be reduced? Let us illustrate this by taking the 2–4 cell in Table 1.39; that is, the cell whose assignment is the amount shipped from f_2 to w_4. At the moment, $x_{24}=0$ as there is no circled number in the cell. Suppose we set $x_{24}=1$. Naturally, we must make some adjustments to the other allocations as f_2 cannot output another bag —it is already shipping its capacity of 40. What we can do is reduce its shipment to w_3 from 30 to 29; that is, set $x_{23}=29$. Then the output of f_2 is back to 40;

$$x_{21}+x_{22}+x_{23}+x_{24}=0+10+29+1=40 \text{ units.}$$

But w_3 is now receiving only 34 bags: 29 from f_2 and 5 from f_3 which is one less than its demand of 35. To compensate we must increase the shipment of f_3 to w_3 to 6; that is, set $x_{33}=6$. This overstretches the output of f_3 by one. So the f_3 to w_4 shipment is reduced by 1, from 20 to 19; that is, we set $x_{34}=19$. This last adjustment is a masterstroke as it simultaneously puts right the output of f_3 (which must be 25) and the demand of w_4 (which must be 20). The final allocation is given in Table 1.52. Each demand and each output is what it should be.

Table 1.52

		Warehouses				
		w_1	w_2	w_3	w_4	Outputs
Factories	f_1	6 (30)	5 (10)	7	9	40
	f_2	3	2 (10)	4 (29)	1 (1)	40
	f_3	7	3	9 (6)	5 (19)	25
Demands		30	20	35	20	

Let us now discover whether or not this slight change has reduced the total cost. The difference in cost between the solutions in Tables 1.39 and 1.52 is as follows:

$+1$ for the bag shipped from f_2 to w_4,
-4 for one bag less shipped from f_2 to w_3,
$+9$ for the extra bag shipped from f_3 to w_3,
-5 for one bag less shipped from f_3 to w_4.
$+1$

What we have found is that an allocation of one unit to the 2–4 cell costs one extra unit and thus is not worthwhile. We remind ourselves of this fact by writing the difference in total cost in the right-hand bottom corner for each cell. The 2–4 cell has a $+1$ there in Table 1.39. We now evaluate the unit costs for all the other empty cells. *The evaluation for each cell is made on the assumption that only it and no other empty cell will be filled.* The changes in the total cost are displayed in the empty cells in Table 1.39. The evaluation of each total cost change is carried out in the same manner as that for the 2–4 cell. In each case we have to trace out a "stepping-stone" circuit of circled numbers which includes the empty cell under consideration. The circuit for cell 2–4 is (2,4), (2,3), (3,3), (3,4), and then back to (2,4). The circuit for each empty cell $i-j$ contains only one empty cell, namely, $i-j$ itself. Some of the circuits for the empty cells in Table 1.39 are longer than the simple one of four cells for cell 2–4:

Empty cell	Circuit
(1–3)	⟨(1–3), (2–3), (2–2), (1–2), (1–3)⟩
(1–4)	⟨(1–4), (3–4), (3–3), (2–3), (2–2), (1–2), (1–4)⟩
(2–1)	⟨(2–1), (1–1), (1–2), (2–2), (2–1)⟩
(2–4)	⟨(2–4), (2–3), (3–3), (3–4), (2–4)⟩
(3–1)	⟨(3–1), (1–1), (1–2), (2–2), (2–3), (3–3), (3–1)⟩
(3–2)	⟨(3–2), (2–2), (2–3), (3–3), (3–2)⟩.

It does not matter whether the circuit is traversed in the clockwise or the counterclockwise direction. The change in total cost for each empty cell is arrived at by alternately adding and subtracting the unit costs of the cells on the circuit as it is traced, starting by adding the unit cost of the empty cell. We have already done this for cell 2–4:

Circuit for the 2–4 cell: ⟨(2–4), (2–3), (3–3), (3–4)⟩

Change in total cost: $1 \quad -4 \quad +9 \quad -5 = 1.$

(Of course we add the cost of the empty cell only once!) Here is another sample calculation:

Circuit for cell 1–4: ⟨(1–4), (3–4), (3–3), (2–3), (2–2), (1–2)⟩

Change in total cost: $9 \quad -5 \quad +9 \quad -4 \quad +2 \quad -5 = 6.$

Only empty cells with negative changes in the total cost will lead to an improved solution. On examining the changes in total cost we see that the most negative is that of the 3–2 cell in Table 1.39, which is -4. This means that for every unit we assign to that cell the total cost decreases by four units. Naturally, we wish to allocate as much as possible to that cell. How much? In calculating the change in total cost for the 3–2 cell we see that we must subtract one unit from the 2–2 and the 3–3 cells, and add one unit to the 3–2 and 2–3 cells. We can do this repeatedly to build up the allocation in the 3–2 cell. Eventually, one of the cells 2–2 and 3–3 from which we have been subtracting will become empty. At this point we cannot make a further subtraction as all cells must have nonnegative allocations. The 2–2 and 3–3 cells start off with 10 and 5 bags, respectively. Thus the most that we can subtract is five bags. Hence this is the most that we can allocate to the 3–2 cell. Each unit allocated to the 3–2 cell reduces the total cost by 4. Thus the maximum allocation of 5 to the 3–2 cell is made and the total is reduced by $5 \times 4 = 20$. The original total cost was 515 units. Thus the total cost of the new solution, shown in Table 1.40, is $515 - 20 = 495$ units. The reader should check that this is the correct cost by calculating it from first principles, that is, by multiplying each allocation by its unit cost and then summing.

The stepping-stone method chooses for allocation the empty cell with the most negative change in total cost. There is the argument that maybe it would be more efficient not to allocate to the empty cell with the most-negative-cost change. Instead maybe we should allocate to the empty cell which brings about the largest cost reduction. To explain this point consider the following example. Suppose we have only two empty cells: e_1 and e_2 with negative-cost changes -10 and -5, respectively. The stepping-stone method allocates as much as possible to e_1 as its cost change is more negative. If we can allocate a maximum of only one bag to e_1 then the total cost is reduced by only 10 units. If we can allocate a maximum of three bags to e_2 then the total cost is reduced by $3 \times 5 = 15$ units. Would it be better in the long run to allocate to e_2 rather than e_1 as the total cost is reduced more? Computer scientists have conducted extensive experiments in order to answer this question. It appears that usually it is unwise to allocate to e_2 rather than e_1. This is because, although it produces an inferior solution, the e_1 allocation requires far less calculation—we do not have to calculate the maximum allocation for each empty cell, only for the one with the most-negative-cost change.

We have now illustrated one complete iteration of the stepping-stone method. We now repeat this iteration on Table 1.40. The method proceeds in this way, generating solution after solution until it produces a table in which no empty cell has a negative-cost change. This table then represents an optimal solution and the method is terminated.

Applying the method to our example problem produces Tables 1.39, 1.40, 1.41(a), and 1.41(b). Table 1.41(b) displays an optimal solution:

$$x_{11}=30, \quad x_{32}=20, \quad x_{13}=10,$$
$$x_{23}=25, \quad x_{24}=15, \quad \text{and} \quad x_{34}=5.$$

The total cost of the solution is 450 units.

It can be seen that in these tables there are empty cells with zero total cost changes. If an allocation is made to such a cell a new feasible solution is produced, which has the same total cost as the present one. We can see that the optimal solution in Table 1.41(b) has such a cell, namely, 2–1. If we allocate as much as we can to it we can produce an alternative optimal solution, shown in Table 1.41(c).

There are interesting phenomena which occur in the calculation of the cost change of the cell 2–1 in Table 1.41(a). Let us trace the circuit of "stepping stones" for this cell. It begins with (2,1) itself of course and then goes upwards to (1,1). We then go to (1,2), down to (3,2), across to (3,4), up to (2,4), and back to (2,1), skipping over (2,3). We skip over (2,3) because the f_2 row and the w_3 column are already "balanced" again by this stage. That is, the modified solution:

$$x_{12}=1, \quad x_{11}=29, \quad x_{12}=11$$
$$x_{32}=9, \quad x_{34}=16, \quad \text{and} \quad x_{24}=4,$$

already allocated the correct output and demand to each factory and warehouse, respectively, including f_2 and w_3. To make it easy to decide which cells to use in the circuit and which to skip over we use the following rule: The circuit has exactly none or two cells from each row and from each column in the table. If we start tracing from the empty cell, this rule helps to define uniquely the circuit.

There is a second point to be made about this circuit $\langle(2,1), (1,1), (1,2), (3,2), (3,4), (2,4), (2,1)\rangle$. It crosses over itself. This should not be a cause for alarm. The crossing over has no effect on the calculation and is quite a common occurrence.

Consider Table 1.42(a). Once the initial solution has been found using the least-cost method it is apparent that only five cells have positive allocations.

It is impossible to use these five cells alone to devise circuits for any of the cells except 3–1. Thus we cannot proceed with the stepping-stone method. In the language of linear programming, we have a degenerate feasible solution. That is, at least one basic variable has value zero. In a transportation problem with n factories and m warehouses there are $m+n$ constraints—one for each factory and one for each warehouse. (We do not count the nonnegativity conditions.) However, one and only one of these constraints is redundant in the sense that it can be deduced by an examination of all the others. This is because we assume that the problem is balanced; that is, total output equals total demand. We have then $n+m-1$ effective constraints. Thus there are $n+m-1$ variables in any basis.

In our present example, $n=3$, and $m=4$ so that we should have six basic variables. But the only ones with positive values are: x_{11}, x_{13}, x_{22}, x_{24}, and x_{33}. We need to declare one more variable basic (with value zero) to bring the basis up to its full complement. We choose the cell of the variable in such a manner as to make it possible to find a circuit for all empty cells. Such cells lie at the intersection of a row and column that each have exactly one positive allocation. Sometimes it is necessary to choose more than one cell to create a basis of $n+m-1$. In our example the 3–4 cell is chosen. (Can you find other cells which make it possible to trace all necessary circuits?) Now x_{34} is set to zero and declared basic. This is indicated in Table 1.42(a) by a circled zero in the 3–4 cell.

The stepping-stone method can now be applied. We find that the cell with the most-negative-cost change is the 3–2 cell. Its circuit is $\langle(3,2), (2,2), (4,2), (4,3), (3,2)\rangle$. The maximum that can be allocated to the 3–2 cell is the minimum of x_{22} and x_{34}, which is zero. Usually the minimum allocation is positive; however, a zero allocation is quite common and need cause no special concern. We treat it in the same way as a basic variable with zero value in the simplex method. Making this allocation produces Table 1.42(b). Naturally it will have the same total cost as the solution in the previous table. However, the next iteration produces Table 1.43(a) from which the degeneracy has disappeared. One more iteration produces the same optimal solution as previously produced. Naturally, it can be modified to produce the alternative optimum.

Table 1.49 displays the initial feasible solution found by the Vogel-approximation method. It is transformed to optimality by the stepping-stone method as shown in Tables 1.50 and 1.51. The same comments about multiple optima made for the other two methods also apply here.

1.2.5. The Transshipment Problem

The transportation model of the last section can be expanded to include the case where locations can act as points of transshipment. This is useful in problems where warehouses not only receive goods but also ship them. The model is often used to analyze the problems of where to locate new warehouses, how many warehouses to build, as well as finding an efficient shipping policy for an existing network. The general problem is called the *transshipment problem*. We now illustrate it with a numerical example.

1.2.5.1. *A Simple Numerical Example*. Consider the numerical example of Section 1.2.1. Suppose that the materials for the cement come from a quarry q and are distributed to factories f_1, f_2, and f_3. Further, the warehouses ship the bags on to two stores, s_1 and s_2. However, f_1 no longer ships its total output of 40 bags. It keeps 10 bags per day for local distribution. Also w_3 keeps 5 of its 35 bags.

However, f_3 supplies an extra eight bags per day for shipment. All the other bags are passed through the system to s_1 and s_2, which demand 40 and 58 bags, respectively. The factory warehouse unit shipping costs remain the same. The quarry must supply 105 bags of materials to f_1, f_2, and f_3, and all these are ultimately shipped to s_1 and s_2 (apart from the 10 kept at f_1 and the five kept at w_3) along with the 8 from f_3. The new distribution network is shown in Fig. 1.4. The unit costs for shipping from q and to s_1 and s_2 are shown next to the appropriate arcs. Above each point is a number indicating the amount it ships on, in excess of what it receives. The point q is termed a *source*; f_1, f_2, f_3, w_1, w_2, w_3, and w_4 are termed transshipment points; and s_1 and s_2 are termed *sinks*. The problem is to supply demands via the transshipment points at minimum total cost.

1.2.5.2. *Transformation to a Transportation Problem*. We now describe how the problem can be transformed into a transportation problem with the same set of feasible solutions. We can then use the stepping-stone method of Section 1.2.4. For our problem, we are going to construct a transportation-problem table, similar to those of Section 1.2.4. Each location capable of shipping bags somewhere else is represented by a row in the table. In our case these are all the locations, except s_1 and s_2. Each location capable of receiving bags is represented by a column in the table. In our case these are all the locations except q. This has been done in Table 1.53, which you should now examine.

The unit shipping cost for each arc is inserted in the upper left-hand corner of the appropriate cell in the usual way. Note that certain direct shipments are impossible, such as q to w_1. In such cases the symbol ∞ is inserted. This represents an arbitrarily large number which ensures that nothing is ever assigned to that cell during the use of the stepping-stone method. Zero unit costs are assigned to the cell at the intersection of the row and column of each transshipment point. Assignments to these cells represent fictitious shipments from a location to itself. This artificial devise is used so that the stepping-stone method can be used. We assign zero costs to these cells in order that they do not affect the total shipping cost.

We now explain the quantities representing capacity and demand, to the right and below the table, respectively. The capacity of q and the demands of s_1 and s_2 are 105, 40, 50 as expected. The difference between the capacity

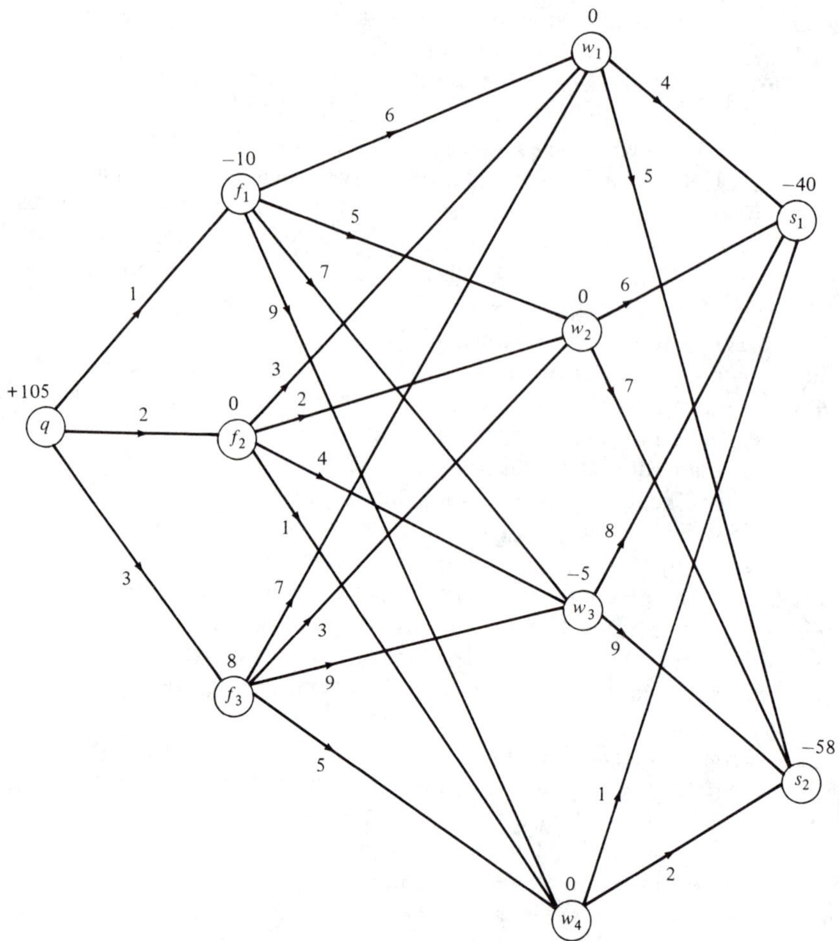

Figure 1.4. The Transshipment Problem.

and demand for each of the other locations is equal to its label in Fig. 1.4. As an example, because w_1 keeps none of its intake, its label is zero, and thus its capacity equals its demand. However, f_1 keeps 10 bags of intake and its label is -10. The demands of the transshipment points have all been set equal to a *buffer stock* of 113, the total number of bags available. The supply of each transshipment point is set equal to the buffer stock plus the supply of the point. (Points f_1 and w_3 have supplies of -10 and -5, respectively.)

Consider the feasible solution given in Table 1.53. (It could not have been obtained by the stepping-stone method as it has too many basic variables.) We can calculate the amount transshipped by each point as follows. For each point with a nonnegative quantity to be supplied, it is the difference between its demand and the quantity shipped to the point. For example, f_2 has zero to be supplied. The number of bags shipped through f_2 equals

Table 1.53

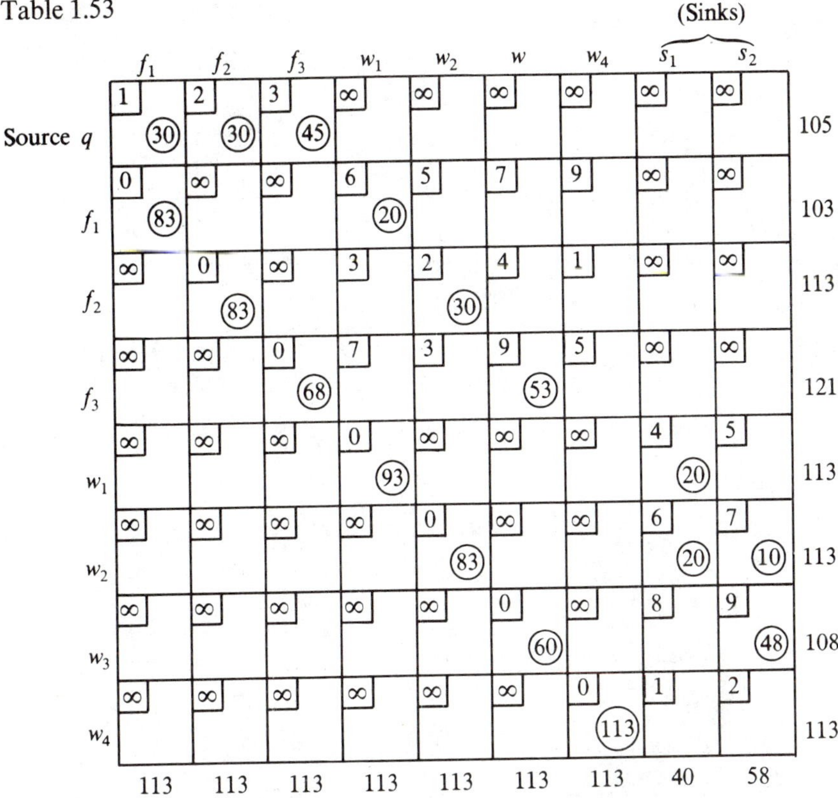

	f_1	f_2	f_3	w_1	w_2	w	w_4	s_1 (Sinks)	s_2 (Sinks)	
Source q	1 (30)	2 (30)	3 (45)	∞	∞	∞	∞	∞	∞	105
f_1	0 (83)	∞	∞	6 (20)	5	7	9	∞	∞	103
f_2	∞	0 (83)	∞	3	2 (30)	4	1	∞	∞	113
f_3	∞	∞	0 (68)	7	3	9 (53)	5	∞	∞	121
w_1	∞	∞	∞	0 (93)	∞	∞	∞	4 (20)	5	113
w_2	∞	∞	∞	∞	0 (83)	∞	∞	6 (20)	7 (10)	113
w_3	∞	∞	∞	∞	∞	0 (60)	∞	8	9 (48)	108
w_4	∞	∞	∞	∞	∞	∞	0 (113)	1	2	113
	113	113	113	113	113	113	113	40	58	

$113-83=30$. For each point with a positive quantity to be supplied, it is the difference between the amount it supplies and the amount shipped to the point. For example, f_1 has 10 bags to be supplied. The number of bags shipped through f_1 equals $103-83=20$.

Consider the basic feasible solution shown in Table 1.54. Beginning with the first row, we can see that q supplies the bare minimum of 10 bags to f_1 and the rest of its supply to f_3. As f_1 has a demand of 10 bags, it cannot ship anything out. This can be verified by calculating the outward shipment from f_1 as $103-103=0$. The 95 bags shipped to f_3 are then boosted by the output of f_3 and the 103 bags are shipped to w_3. The number of bags transshipped at f_3 can be calculated as $113-18=95$. The number transshipped at f_2 is zero which can be verified by calculating $113-113=0$. The 103 bags received at w_3 are all transshipped to supply the demands of s_1 and s_2, except the five-bag demand of w_3 itself. The number transshipped at w_3 can be verified as $108-10=98$. The quantities transshipped at w_1, w_2, and w_4 can all be found to be zero.

1.2.5.3. *General Transformation into a Transportation Problem.* The transportation problem for a transshipment problem is constructed in Table 1.54.

Table 1.54

	f_1	f_2	f_3	w_1	w_2	w_3	w_4	s_1	s_2	
q	10		95							105
f_1	103									103
f_2		113								113
f_3			18			103				121
w_1				113						113
w_2					113					113
w_3						10		40	58	108
w_4							113			113
	113	113	113	113	113	113	113	40	58	

Steps

(1) Assign a row i for each source i. Set b_i = the capacity of the source i.
(2) Assign a column j for each sink j. Set a_j = the demand of the sink j.
(3) Assign a row and a column for every transshipment point k. Set a_k = the total capacity of all sources. Set $b_k = a_k + t_k$, where t_k = the net output of point k. (If k has a positive capacity for supply, t_k is positive. If k has a positive demand, t_k is negative.)
(4) Set c_{ij} the unit shipping cost from point i to point j, to be ∞, an arbitrarily large number, if it is not possible to supply point j directly from point i. Set $c_{ii} = 0$ if i is a transshipment point. (All other c_{ij}'s must be given.)

1.3. The Assignment Problem

The assignment problem is a special kind of transportation problem in which the number of "factories" equals the number of "warehouses" and each output and demand is exactly one unit. Because of this structure it can be solved more efficiently by a specialized algorithm, called the "Hungarian method," than by using the stepping-stone method. The problem is usually described in terms of matching n objects with n other objects in a one-to-one fashion. We do this now.

1.3.1. A Simple Assignment Problem

The Won't Wear Out carpet company manufactures 10 carpets. As each type of carpet comes off the loom it is examined for faults and repaired by hand sewing called *picking*. There is one picking board and one picker for each carpet. The company has 10 workers who have volunteered for picking. People work at different rates on the different carpets because of individual preferences over the patterns, the design, board size, and factory location. Naturally the company wants to make the most efficient assignment possible of the 10 workers to the 10 boards. It devises a standardized task for each worker to perform on each carpet and notes the completion times in minutes. These times are displayed in Table 1.55, where row i represents worker w_i and column j represents carpet c_j. The problem is to match each worker with a unique carpet so that the sum of the completion times of the workers on carpets to which they are matched is minimized. We now define some decision variables in order to build a mathematical model of this problem. Let

$$x_{ij}\begin{cases} =1 & \text{if } w_i \text{ is assigned to } c_j, \\ =0 & \text{otherwise.} \end{cases}$$

We begin by constructing the objective function. What is the total completion time of an assignment defined by a set of values for the x_{ij} variables? If w_1 is assigned to c_1, then the time of 3 minutes should be included because this is the entry in the row 1, column 1 position of Table 1.55. If w_1 is not

Table 1.55

		Carpets										
		1	2	3	4	5	6	7	8	9	10	
	1	3	3	10	9	5	2	11	2	11	5	(2)
	2	6	2	7	11	4	10	4	4	5	4	(2)
	3	9	7	9	10	4	4	5	5	4	5	(4)
	4	8	6	7	8	8	8	10	6	3	9	(3)
	5	7	2	8	6	10	9	6	6	11	10	(2)
Workers	6	5	11	3	6	10	3	6	7	2	10	(2)
	7	4	11	11	5	9	11	7	9	10	11	(4)
	8	11	10	5	4	11	4	7	8	7	3	(3)
	9	11	5	5	3	2	5	7	10	7	3	(2)
	10	10	4	5	2	11	6	11	7	8	2	(2)

assigned to c_1, then we should not include this time of 3 minutes. From the definition of the x_{ij}'s it can be seen that we can include the term $3x_{11}$ in the objective function to cover both eventualities. Indeed we can merely sum the products of the entries of Table 1.55 and their corresponding x_{ij}. The objective is to

$$\text{Minimize } 3x_{11}+3x_{12}+10x_{13}+\cdots+2x_{10,10}.$$

We now identify the constraints. Worker w_1 must be assigned exactly one of the 10 carpets. Suppose it is c_3. Then $x_{13}=1$ and $x_{ij}=0$, where $j=1,2,4,5,\ldots,10$.

$$\therefore x_{11}+x_{12}+x_{13}+\cdots+x_{1,10}=1. \tag{1.38}$$

Indeed no matter which carpet is assigned to w_1, (1.38) will hold. We can define a constraint like (1.38) for each worker i:

$$x_{i1}+x_{i2}+x_{i3}+\cdots+x_{i,10}=1, \qquad i=1,2,\ldots,10.$$

Also c_1 must be assigned exactly one of the 10 workers. If it is w_4, we have $x_{41}=1$ and $x_{i1}=0$, where $i=1,2,3,5,\ldots,10$.

$$\therefore x_{11}+x_{21}+x_{31}+\cdots+x_{10,1}=1. \tag{1.39}$$

Indeed no matter which worker is assigned to c_1, (1.39) will hold. We can now define a constraint like (1.39) for each carpet j:

$$x_{1j}+x_{2j}+x_{3j}+\cdots+x_{10,j}=1, \qquad i=1,2,\ldots,10.$$

Further each variable must be equal to either zero or one:

$$x_{ij}=0 \text{ or } 1, \qquad i=1,2,\ldots,10, \; j=1,2,\ldots,10.$$

The complete mathematical model is

$$\text{Minimize } 3x_{11}+3x_{12}+10x_{13}+\cdots+2x_{10,10}$$

subject to

$$\begin{aligned} x_{i1}+x_{i2}+x_{i3}+\cdots+x_{i,10}&=1, && i=1,2,\ldots,10,\\ x_{1j}+x_{2j}+x_{3j}+\cdots+x_{10,j}&=1, && j=1,2,\ldots,10,\\ x_{ij}&=0 \text{ or } 1, && i=1,2,\ldots,10, \; j=1,2,\ldots,10. \end{aligned}$$

1.3.2. The General Formulation

The problem introduced in the last section is an example of the *assignment problem*. We now introduce a general model. Let

$n=$ the number of workers (which is also assumed to be the number of carpets),

$t_{ij}=$ the completion time of w_i on c_j.

The problem is to

$$\text{Minimize} \sum_{i=1}^{n} \sum_{j=1}^{n} t_{ij} x_{ij}$$

subject to

$$\sum_{j=1}^{n} x_{ij} = 1, \qquad i = 1, 2, \ldots, n, \tag{1.40}$$

$$\sum_{i=1}^{n} x_{ij} = 1, \qquad j = 1, 2, \ldots, n, \tag{1.41}$$

$$x_{ij} = 0 \text{ or } 1, \; i = 1, 2, \ldots, n, \; j = 1, 2, \ldots, n.$$

Compare this formulation with the general transportation problem formulation (1.37). It can be seen that this one is a special case of the former where:

(1) $m = n$.
(2) $b_j = 1, \; j = 1, 2, \ldots, n$.
(3) $a_i = 1, \; i = 1, 2, \ldots, n$.

The constraints are equations rather than inequalities. This is because once the transportation problem is balanced by the introduction of a dummy (if necessary) and total output equals total demand, its inequalities become equations.

We have assumed that the number of workers equals the number of carpets. If this is not the case we must introduce sufficient dummy workers or dummy carpets in order to make the two numbers equal. This is done in much the same way as imbalance in the transportation problem is corrected (see Section 1.2.1). All times concerned with a dummy are once again defined to be zero. As an example consider a three-worker, five-carpet-assignment problem with completion times

		Carpets				
		1	2	3	4	5
	1	6	8	7	18	6
Workers	2	9	4	3	4	12
	3	12	5	2	10	8

We introduce two extra dummy workers, w_4 and w_5, with zero completion times for all carpets:

		Carpets				
		1	2	3	4	5
	1	6	8	7	18	6
	2	9	4	3	4	12
Workers	3	12	5	2	10	8
	4	0	0	0	0	0
	5	0	0	0	0	0

A shortage of carpets can be handled in the same way. We now introduce a method of solution for the balanced assignment problem.

1.3.3. The Hungarian Method

We introduce the method via our example problem. Please examine Table 1.55. In essence what we wish to do is to circle 10 numbers representing 10 times in this table. Each circled number corresponds to an assignment—the assignment of the worker of its row to the carpet of its column. The circled numbers must be such that (i) there is exactly one circled number in each row (each worker must be given exactly one carpet) and (ii) exactly one circled number is in each column (each carpet must be given to exactly one worker). Among all possible sets of 10 circled numbers, we seek the set with the minimum sum. Now consider Table 1.58. It contains a new set of times. In this case it is quite easy to detect an optimal solution, such as the one shown. If we can find a set of zeros obeying requirements (i) and (ii) above, it must be optimal as we assume that all the times are nonnegative. So our aim is to somehow transform the matrix of times [so that we can find a set of zeros obeying (i) and (ii)], *without altering the set of optimal solutions to the original problem*. The total time of the final assignment represented by the set of zeros will of course be zero. Thus the total time of the final assignment will be less than the actual total time for the original matrix. However, the plan is to ensure that the actual assignment itself is optimal for the original problem. We now show how this can be done.

Suppose we subtract two units from every entry in Table 1.55. We would then produce some zeros. Because the relative values of the entries remain unaltered, a minimal solution for the modified matrix will be minimal for the original one. We can think of the 2 minutes subtracted as a time which all workers must take, such as for setting up. This can be neglected as it plays no part in the optimization. We now define how to make the correct subtractions in order to arrive at a desired set of zeros. The method is iterative in the sense that it progressively defines a series of matrices until a solution of zeros can be identified. Let the original matrix of times be denoted by **T**.

General Outline of the Hungarian Method

(1) For each row in **T** subtract the smallest entry in the row from all entries in the row.
(2) For each column with all positive entries subtract the smallest entry in the column from all entries in the column.

(3) Check to see whether a feasible solution of all zero entries can be identified. If so it represents an optimal solution and the method is terminated. If not go to step (4).
(4) Redistribute the zeros in the matrix and go back to step (3).

We shall now make these steps more definite. First, we must show that steps (1) and (2) do not alter the set of optimal solutions. That is, any optimal solution for **T** will still be optimal for the matrix created by steps (1) and (2). Let

$d_i =$ the amount subtracted from the ith row, $\qquad i=1,2,\ldots,m;$

$e_j =$ the amount subtracted from the jth row column, $\qquad j=1,2,\ldots,m;$

$\mathbf{T}' = \left(t'_{ij}\right)_{m\times m} =$ the matrix created by steps (1) and (2).

Then

$$t'_{ij} = t_{ij} - d_i - e_j.$$

The new objective function equals

$$\begin{aligned}\sum_{i=1}^{m}\sum_{j=1}^{m} t'_{ij}x_{ij}, &\\ &= \sum_{i=1}^{m}\sum_{j=1}^{m}\left(t_{ij}-d_i-e_j\right)x_{ij},\\ &= \sum_{i=1}^{m}\sum_{j=1}^{m} t_{ij}x_{ij} - \sum_{i=1}^{m}\sum_{j=1}^{m} d_i x_{ij} - \sum_{i=1}^{m}\sum_{j=1}^{m} e_j x_{ij},\\ &= \sum_{i=1}^{m}\sum_{j=1}^{m} t_{ij}x_{ij} - \sum_{i=1}^{m} d_i \sum_{j=1}^{m} x_{ij} - \sum_{j=1}^{m} e_j \sum_{i=1}^{m} x_{ij}.\end{aligned}$$

By (1.40) and (1.41):

$$\sum_{i=1}^{m}\sum_{j=1}^{m} t'_{ij}x_{ij} = \sum_{i=1}^{m}\sum_{j=1}^{m} t_{ij}x_{ij} - \sum_{i=1}^{m} d_i - \sum_{j=1}^{m} e_j.$$

Thus it can be seen that the original and new objective function values differ only in the total amount subtracted, which is a constant. Therefore the problems defined by **T** and **T′** have identical optimal solutions.

We now define an efficient way to perform step (3):

(3i) Identify a row or column in **T′** with exactly one zero. If there is no such row or column, choose the row or column with the smallest number of zeros. Ties are settled arbitrarily. Identify a zero in the row (column) selected. If it is a row that has been identified, draw a vertical line through this zero. If a column has been identified, draw a horizontal line through this zero.

(3ii) Repeat step (3i), ignoring any zeros with a line through it, until each zero has at least one line through it.

(3iii) If exactly m lines have been drawn, an optimal set of zeros is present. The zeros identified for the drawing of lines comprise a solution—it may not be unique. If less than m lines have been drawn, an optimal set of zeros is not yet at hand. In this case go to step (4).

Steps (3i), (3ii), and (3iii) will be made clear via the numerical example shortly. We now come to step (4). The zeros are redistributed by taking advantage of Theorem 1.3.

Theorem 1.3 (Konig's Theorem). *Let the entries of a matrix* $\mathbf{T}$ *be divided into exactly two classes by property* P. *Then the minimum number of straight lines that contain all the entries with property* P *is equal to the maximum number of entries with property* P, *no two on the same line.*

We take advantage of this theorem where in our case property P is to be equal to zero. Step (4) in expanded form is:

(4i) Subtract the minimum entry in $\mathbf{T}'$ with no line through it from each entry with no line through it.

(4ii) Add the number subtracted in step (4i) to each entry in $\mathbf{T}'$, which has two lines (both horizontal and vertical) through it.

(4iii) Remove all lines.

We shall now illustrate the Hungarian method by applying it to the numerical example. We begin with step (1) and subtract the minimum entry in each row from the row. The numbers subtracted are shown in parentheses on the right. This produces Table 1.56. We then apply step (2) and find that only two columns now have all positive entries—columns 3 and 7. Their minimum entries are subtracted (shown in parentheses) which produces Table 1.57. We now apply step (3) and draw the minimum possible number of lines needed to cross out all the zeros. Only nine lines are required which is less than the 10 required for success. So we move to step (4). The minimum uncrossed number in Table 1.57 is 1. It is subtracted from all uncrossed numbers and added to all doubly crossed numbers. The lines are then removed. This produces Table 1.58. Returning to step (3) it turns out that we now need 10 lines to cross out all the zeros. [If we had needed less than 10 we would have returned to step (4)]. An optimal solution is at hand. We can find it by identifying the zeros through which lines are drawn. There are a number of choices and thus there is more than one optimal solution. One of the optimal solutions is circled in Table 1.58. (Can you find all the others?) This solution is

$$x_{18} = x_{27} = x_{36} = x_{49} = x_{52} = x_{63} = x_{71} = x_{8,10} = x_{95} = x_{10,4} = 1.$$

Table 1.56

1	1	8	7	3	0	9	0	9	3
4	0	5	9	2	8	2	2	3	2
5	3	5	6	0	0	1	1	0	1
5	3	4	5	5	5	7	3	0	6
5	0	6	4	8	7	4	4	9	8
3	9	1	4	8	1	4	5	0	8
0	7	7	1	5	7	3	5	6	7
8	7	2	1	8	1	4	5	4	0
9	3	3	1	0	3	5	8	5	1
8	2	3	0	9	4	9	5	6	0
		(1)				(1)			

Table 1.57

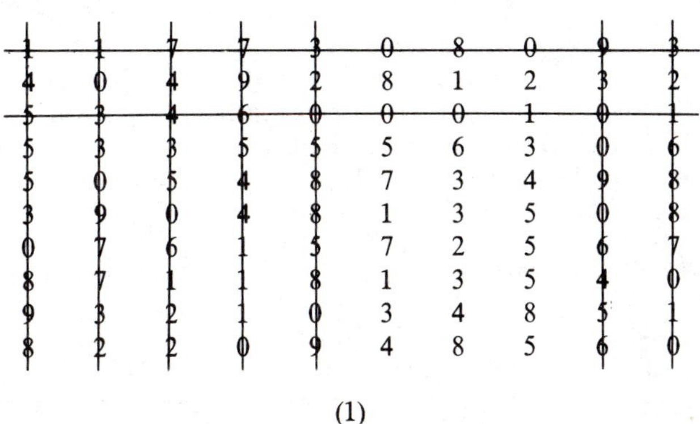

1	1	7	7	3	0	8	0	9	3
4	0	4	9	2	8	1	2	3	2
5	3	4	6	0	0	0	1	0	1
5	3	3	5	5	5	6	3	0	6
5	0	5	4	8	7	3	4	9	8
3	9	0	4	8	1	3	5	0	8
0	7	6	1	5	7	2	5	6	7
8	7	1	1	8	1	3	5	4	0
9	3	2	1	0	3	4	8	5	1
8	2	2	0	9	4	8	5	6	0
						(1)			

Table 1.58

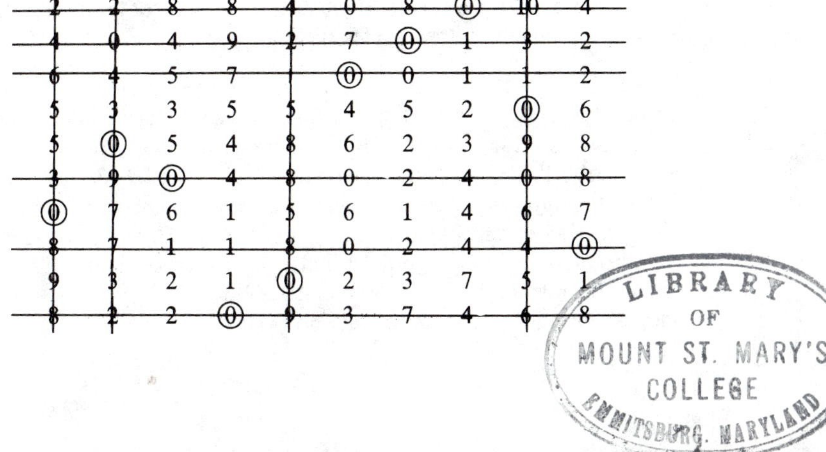

2	2	8	8	4	0	8	(0)	10	4
4	0	4	9	2	7	(0)	1	3	2
6	4	5	7	1	(0)	0	1	1	2
5	3	3	5	5	4	5	2	(0)	6
5	(0)	5	4	8	6	2	3	9	8
3	9	(0)	4	8	0	2	4	0	8
(0)	7	6	1	5	6	1	4	6	7
8	7	1	1	8	0	2	4	4	(0)
9	3	2	1	(0)	2	3	7	5	1
8	2	2	(0)	9	3	7	4	6	8

All other x_{ij}'s are zero. Thus an optimal assignment is:

Worker:	1	2	3	4	5	6	7	8	9	10
Carpet:	8	7	6	9	2	3	1	10	5	4.

The total completion time of this assignment can be calculated by referring to the original matrix:

$$2+4+4+3+2+3+4+3+2+2=29 \text{ min.}$$

A useful check can be made on this calculation. The total completion time of an optimal assignment is equal to the total amount subtracted from the original matrix. So adding the number in parentheses, plus the 1 subtracted in step (4):

$$2+2+4+3+2+2+4+3+2+2+1+1+1=29 \text{ min.}$$

1.4. Exercises

1. A shop produces two types of cake. Type 1 requires 5 lb of flour F_1 and 4 lb of flour F_2 per batch. Type 2 requires 4 and 6 lb of F_1 and F_2 per batch, respectively. The profit is \$1 and \$2 for a batch of type 1 and 2, respectively. There is 200 and 240 lb of F_1 and F_2 available. How many batches of each type of cake should be produced in order to maximize profit? Solve this problem both graphically and by the simplex method.

2. A brewery makes three beers: B_1, B_2, and B_3. The amounts (in pounds) of sugar, malt, and hops required for each are shown below:

Beer	Profit/barrel	Sugar	Malt	Hops
B_1	\$30	1	2	1
B_2	\$20	2	1	1
B_3	\$30	3	1	1

Every hour, 6, 4, and 3 lb of sugar, malt, and hops can be fed into the process. How much of each beer should be made per hour in order to maximize profit? Solve this problem by the simplex method.

3. Create and solve the dual for the following problem. Find the optimal solution for the problem by interpreting the optimal dual tableau. An engineering plant makes two products: P_1 and P_2 at a profit of \$500 and \$200 per pound, respectively. P_1 (P_2) requires per pound, 2, 5, 5, and 2 (3, 2, 3, and 1) lb of nickel (N) chromium (C), germanium (G), and magnesium (M), respectively. The plant can obtain 7, 11, 10, and 6 lb of N, C, G, and M, respectively, per day. Products P_1 and P_2 require 1 and 2 hr of furnace time daily. The furnace can be operated for only 6 hr per day. What production policy should be adopted in order to maximize daily profit?

4. Solve the following unbalanced transportation problem:

	Warehouses				Outputs
Factories	7	5	4	3	7
	5	4	4	3	8
	6	5	6	7	11
	3	4	7	9	5
Demands	6	12	5	7	

5. Solve the following unbalanced assignment problem:

	Machines				
Workers	8	3	7	7	12
	3	4	1	7	19
	4	5	6	1	20
	1	7	5	5	14
	6	9	4	12	13
	9	2	3	14	12

6. Form the dual of the following problem.

$$\text{Maximize } 2y_1 + 4y_2 + y_3 + 3y_4$$

subject to

$$\begin{aligned} 2y_1 - y_2 + y_3 + 2y &\leq 6, \\ 2y_2 \qquad - y_4 &\leq 1, \\ y_1 + y_2 \qquad + 2y_4 &\leq 4, \\ 2y_1 + 2y_2 + 2y_3 + y_4 &\leq 9, \\ y_1, \ldots, y_4 &\geq 0. \end{aligned}$$

(a) Solve the dual problem by the dual simplex method.
(b) Solve the dual problem by the regular simplex.
(c) Compare the computational effort required in (a) and (b).
(d) From (a) and (b) deduce the solution to the primal problem.

*7. Prove that if a linear programming problem has an optimal solution with infinite value then its dual does not have any feasible solutions.

8. Apply the stepping-stone method to convert the solution given in Table 1.53 into a table depicting the optimal solution.

*9. Describe how the solution given in Table 1.52 can be converted into a basic feasible solution.

10. Carry out the conversion required in problem 9 and convert the resulting solution into the optimal solution.

*11. Describe how the northwest-corner method can be modified to produce an initial basic feasible solution for the transshipment problem which does not involve any prohibited arcs (i.e., no ∞ costs are involved).

*12. Repeat problem 10 for the least-cost method.

*13. Repeat problem 11 for the Vogel-approximation method.

CHAPTER 2

Solution Techniques

This chapter describes some of the basic solution techniques for combinatorial optimization problems—integer programming, dynamic programming, and heuristic problem solving. It begins with integer programming (IP), that is, the optimization of a linear function subject to a number of linear constraints, where the variables of the function must be integers. The elementary notions of two broad approaches to integer programming—enumerative techniques and cutting planes, are covered.

Dynamic programming (DP) is an approach for solving problems in which decisions are to be made in stages. It is easy to think of real-world situations that are multistage decision problems. The investment of funds over a period of time and construction projects are two of the multitude of common examples. Dynamic programming also uses the philosophy of implicit enumeration as does one of the integer-programming approaches. The scope of dynamic programming is very vast indeed, so only a modest introduction to the elementary DP techniques is attempted here. The interested reader is referred to the Further Reading section at the end of the book for a guide to specialist DP books.

The two previously mentioned techniques are designed to produce global extrema for the problems to which they are applied. Unfortunately, many real-world problems are so large and difficult that these methods cannot achieve this extrema efficiently due to their large storage or computational-time requirements. In this case a more modest aim is in order. Instead of requiring an absolute optimum, it is often more realistic to design a solution procedure which will produce, in reasonable computing time, solutions which are relatively close to the optimum. Such procedures are called *heuristics*.

An example of an heuristic was given in Chapter 0 for the shortest Hamiltonian Path problem. There is a wide variety of uses for heuristics. They may be used to attempt to solve well-defined, large, difficult problems, or ill-defined problems with imprecise constraints and multiple-objective criteria. They are also used as subroutines within larger procedures, to construct initial solutions, and to improve on existing solutions. There is something of an art, rather than a science, to heuristic design as will be seen later in this chapter.

2.1. Integer Programming

The Quick as Lightning freight company has a problem. It has entered into a long-term contract with a corporation which makes three types of machine parts: A, B, and C, which are then stored in very large crates. The contract requires the company to ship periodically these crates from the manufacturing plant to a distant warehouse. The company decides to use its only aircraft (a DC3) to its best advantage and carry the rest of the crates by truck. The question is, Which goods should be airfreighted? The volume of one crate of A, B, and C is two, six, and three units, respectively, the DC3 having eight units of volume capacity. The weight of one crate of A, B, and C is five, four, and four units, respectively, the DC3 having seven units of weight capacity. The value of one crate of A, B, and C is 60, 10, and 10 units, respectively and due to the nature of the company's insurance policy the plane must not carry goods worth more than 120 units. Further, the revenue gained by transporting one crate of A, B, and C is three, five, and four units, respectively. The company decides to calculate how many crates each of A, B, and C to carry in order to maximize revenue.

In order to settle this let

$$x_1 = \text{the number of crates of type A carried};$$

$$x_2 = \text{the number of crates of type B carried};$$

$$x_3 = \text{the number of crates of type C carried}.$$

Thus if x_1 crates of type A are carried a revenue of $3x_1$ is produced. Similarly, revenues of $5x_2$ and $4x_3$ will be produced for types B and C, respectively. The total revenue for a policy of carrying x_1, x_2, and x_3 of A, B, and C is

$$3x_1 + 5x_2 + 4x_3.$$

As one crate of type A occupies two units of volume, x_1 crates will occupy $2x_1$ units. We can make a similar deduction for B and C. Hence the x_1, x_2, x_3 policy requires $2x_1 + 6x_2 + 3x_3$ units of volume. But as only eight units are available we must have

$$2x_1 + 6x_2 + 3x_3 \leq 8.$$

Similar reasoning leads to constraints for weight:

$$5x_1+4x_2+4x_3 \leq 7,$$

and for value after dividing by 10:

$$6x_1+x_2+x_3 \leq 12.$$

The problem can now be formally defined as

$$\text{Maximize } 3x_1+5x_2+4x_3 \quad (=Z) \tag{2.1}$$

subject to

$$2x_1+6x_2+3x_3 \leq 8, \tag{2.2}$$

$$5x_1+4x_2+4x_3 \leq 7, \tag{2.3}$$

$$6x_1+\ x_2+\ x_3 \leq 12, \tag{2.4}$$

$$x_1, x_2, x_3, \text{ nonnegative integers.} \tag{2.5}$$

It is the last constraint (2.5) that distinguishes the problem from one of linear programming. Before looking for methods which will solve this problem, let us present the general formulation of which our airfreight problem is a special case.

In terms of the definitions given in Section 1.1, S is a set of n-dimensional real vectors $\mathbf{X}$, where the following restrictions apply. There are given two sets of n-variable, real-valued functions g_j, where $j=1,2,\ldots,m$, and h_i, where $i=1,2,\ldots,k$, such that

$$g_j(\mathbf{X})=0, \qquad j=1,2,\ldots,m,$$

and

$$h_i(\mathbf{X}) \leq 0, \qquad i=1,2,\ldots,k,$$

for any $\mathbf{X} \in S$. Further there is a given q, $1 \leq q \leq n$, such that $x_1, x_2, \ldots, x_q$ are integers, where $\mathbf{X}=(x_1, x_2, \ldots, x_q, x_{q+1}, \ldots, x_n)$. Stated formally, the *general integer programming problem* is

$$\text{Maximize } f(\mathbf{X})$$

subject to

$$g_j(\mathbf{X})=0, \qquad j=1,2,\ldots,m,$$

$$h_i(\mathbf{X}) \leq 0, \qquad i=1,2,\ldots,k,$$

$$\mathbf{X}=(x_1, x_2, \ldots, x_q, x_{q+1}, \ldots, x_n),$$

where

$$x_1, x_2, \ldots, x_q \text{ are integers for a given } q.$$

As this problem remains essentially unsolved in the general case we confine our attention to a useful simplification. We assume that f and the h_i's are linear, there are no g_j's, and all the variables in $\mathbf{X}$ must be nonnegative. Then the formulation can be expressed in matrix notation as

$$\text{Maximize } \mathbf{CX} \tag{2.6}$$

subject to

$$\mathbf{AX} \leq \mathbf{b}, \tag{2.7}$$

$$\mathbf{X} \geq \mathbf{0}, \tag{2.8}$$

$$x_1, x_2, \ldots, x_q, \text{ integers}, \tag{2.9}$$

where

$$\mathbf{X} = (x_1, x_2, \ldots, x_q, x_{q+1}, \ldots, x_n)^T,$$

$\mathbf{C}$ is a $1 \times n$ real vector,

$\mathbf{b}$ is an $m \times 1$ real vector,

$\mathbf{A}$ is an $m \times n$ real matrix.

If

$$q = n,$$

the problem is termed an *all-integer linear-programming* (IP) *problem.* Indeed our airfreight problem is of this type with

$$\mathbf{C} = (3,5,4),$$

$$\mathbf{b} = \begin{bmatrix} 8 \\ 7 \\ 12 \end{bmatrix},$$

and

$$\mathbf{A} = \begin{pmatrix} 2 & 6 & 3 \\ 5 & 4 & 4 \\ 6 & 1 & 1 \end{pmatrix}.$$

If $1 < q < n$, the problem is termed a *mixed-integer linear-programming problem*. If (2.9) is replaced by

$$x_i = 0 \quad \text{or} \quad 1, \qquad i = 1, 2, \ldots, n,$$

the problem is termed a *zero–one programming problem.*

Because most of the research presented in the literature has been concerned with the integer-linear-programming problem, the word "linear" is often dropped from use here. Having fitted the airfreight problem into the general framework, we now present ways to solve it.

2.1.1. Rounding

One obvious approach to (2.6)–(2.9) is to neglect (2.9) and use the powerful simplex method on the resulting problem. If the solution produced satisfies (2.9) then it must be optimal. If it does not then there are a number of options available. One straightforward strategy is to round the values of noninteger variables either up or down to achieve an integer solution.

Let us explore this idea on the following integer-programming problem:

$$\text{Maximize } x_1 + x_2 \quad (=Z)$$

subject to

$$\begin{aligned} 15x_1 + 12x_2 &\leq 85, \\ 5x_1 \qquad &\geq 11, \\ x_1, \quad x_2 &\geq 0, \\ x_1, \quad x_2, &\text{ integers.} \end{aligned}$$

The linear-programming version of this problem has been solved graphically in Fig. 2.1. It can be seen that the optimal solution is

$$(x_1, x_2) = (\tfrac{11}{5}, \tfrac{13}{3}), \qquad Z = \tfrac{98}{15}.$$

Rounding each of these values up and down produces the four solutions: (2,4), (3,4), (2,5), and (3,5). However, as can be seen, none of these solutions are feasible. Indeed the optimal solutions for the original problem are

$$(x_1, x_2) = (3,3), (4,2), \text{ or } (5,1), \qquad Z = 6.$$

So even on a problem as small as this there may be severe problems with rounding. The process may fail to produce a feasible solution, let alone the

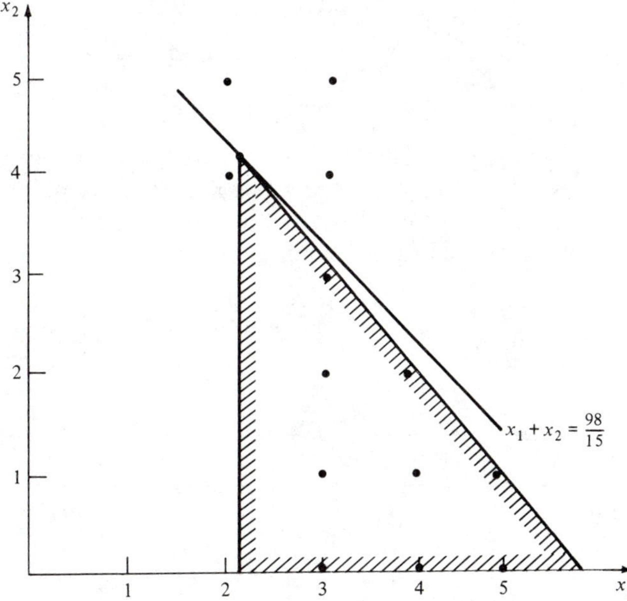

Figure 2.1. The Failure to Obtain an IP Solution by Rounding an LP Solution.

optimal one. Even when a rounded solution is feasible it is sometimes quite different in value, relatively speaking, from the optimum.

Given these problems of rounding we can, of course, use instead the fundamental algorithm. That is, list all feasible solutions and choose the best. We now explore a modified version of this approach.

2.1.2. Branch-and-Bound Enumeration

In this section we unveil a method that guarantees to find an optimal solution (if one exists) to any integer-programming problem. It is an *implicit enumeration* technique and is thus named because it implicitly examines all possible solutions in its search for the optimum. Another example of implicit enumeration—dynamic programming—will be discussed in Section 2.2.

The approach is called "branch-and-bound enumeration" and a variation due to Dakin (1965) is explained first.

2.1.2.1. *Dakin's Method.* In order to explain Dakin's method we use it to solve the airfreight problem [Eqs. (2.1)–(2.5)]:

$$\text{Maximize } 3x_1 + 5x_2 + 4x_3 \quad (= Z)$$

subject to

$$2x_1 + 6x_2 + 3x_3 \leq 8,$$
$$5x_1 + 4x_2 + 4x_3 \leq 7,$$
$$6x_1 + x_2 + x_3 \leq 12,$$
$$x_1, x_2, x_3, \text{ nonnegative integers.}$$

We begin by neglecting the fact that x_1, x_2, and x_3 are required to be integers and solve the problem by the simplex method. Introducing slack variables x_4, x_5, and x_6 into (2.2), (2.3), and (2.4), respectively, Table 2.1 provides us with the optimal LP solution:

$$x_1^* = 0, \qquad x_2^* = \tfrac{11}{12}, \qquad x_3^* = \tfrac{5}{6}, \qquad Z^* = \tfrac{95}{12}.$$

Table 2.1

x_1	x_2	x_3	x_4	x_5	x_6	**b**
$-\frac{7}{12}$	1	0	$\frac{1}{3}$	$-\frac{1}{4}$	0	$\frac{11}{12}$
$\frac{11}{6}$	0	1	$-\frac{1}{3}$	$\frac{1}{2}$	0	$\frac{5}{6}$
$\frac{19}{4}$	0	0	0	$-\frac{1}{4}$	1	$\frac{41}{4}$
$\frac{17}{12}$	0	0	$\frac{1}{3}$	$\frac{3}{4}$	0	$\frac{95}{12}$

If this was a feasible IP solution it would be the optimal IP solution and nothing more would have to be done. As this is not the case, Dakin's method does not round this solution but identifies just one of the variables with a fractional value, say,

$$x_2^* = \tfrac{11}{12}.$$

Taking the integer part of this value of $\frac{11}{12}$, which is zero, we form two constraints:

$$x_2 \leq 0 \tag{2.10}$$

and

$$x_2 \geq 0+1. \tag{2.11}$$

Stated in words, "we form one constraint with the variable less than or equal to the integral part and another with the variable greater than or equal to the integral part plus one." Now as x_2 is constrained to be a nonnegative integer, it must either be 0 or greater than or equal to 1. Suppose two new problems, (I) and (II) are created—one by adding (2.10) to the original problem (2.1)–(2.5) and the other by adding (2.11). Then the optimal IP solution to (2.1)–(2.5) must be optimal for exactly one of (I) and (II).

As x_2 denotes the number of type B crates that are accepted, all we are doing in this particular case is listing the two possibilities: (i) no B crates are flown and (ii) at least one B crate is flown. The two new LP problems are:

(I)

Maximize $3x_1 + 5x_2 + 4x_3$

subject to

$$\begin{aligned} 2x_1 + 6x_2 + 3x_3 &\leq 8, \\ 5x_1 + 4x_2 + 4x_3 &\leq 7, \\ 6x_1 + x_2 + x_3 &\leq 1, \\ x_2 &\leq 0, \\ x_1, x_2, x_3 &\geq 0. \end{aligned}$$

(II)

Maximize $3x_1 + 5x_2 + 4x_3$

subject to

$$\begin{aligned} 2x_1 + 6x_2 + 3x_3 &\leq 8, \\ 5x_1 + 4x_2 + 4x_3 &\leq 7, \\ 6x_1 + x_2 + x_3 &\leq 12, \\ x_2 &\geq 1, \\ x_1, x_2, x_3 &\geq 0. \end{aligned}$$

It has turned out in our given instance that our nonintegral value ($\frac{11}{12}$) is fractional. Thus the integral part is zero. Thus (2.10), together with the nonnegativity condition on x_2, implies that in any feasible solution to (I),

$$x_2 = 0.$$

So we could simply remove x_2 from the original formulation and solve a new LP problem in order to find the optimum to (I). However, in general, the integral part will be positive and so this approach is inappropriate. Even more importantly, we can use the dual simplex method (see Section 1.1.7) to arrive at the optimum to the new LP problems very efficiently.

Table 2.2

x_1	x_2	x_3	x_4	x_5	x_6	x_7	rhs
$-\frac{7}{12}$	1	0	$\frac{1}{3}$	$-\frac{1}{4}$	0	0	$\frac{11}{12}$
$\frac{11}{6}$	0	1	$-\frac{1}{3}$	$\frac{1}{2}$	0	0	$\frac{5}{6}$
$\frac{19}{4}$	0	0	0	$-\frac{1}{4}$	1	0	$\frac{41}{4}$
0	1	0	0	0	0	1	0
$\frac{17}{12}$	0	0	$\frac{1}{3}$	$\frac{3}{4}$	0	0	$\frac{95}{12}$

We now apply the dual simplex method to Table 2.2 in order to incorporate (2.10); that is,

$$x_2 \leq 0$$

becomes, with slack variable x_7,

$$x_2 + x_7 = 0.$$

In canonical form, we arrive at Tables 2.3 and 2.4. This has solution:

$$x_1^* = 0,$$

$$x_2^* = 0 \quad (\text{as expected as } x_2 \leq 0 \text{ and } x_2 \geq 0),$$

$$x_3^* = \tfrac{7}{4},$$

$$Z^* = 7.$$

We can apply the dual simplex method in a similar manner to (II) and

Table 2.3

x_1	x_2	x_3	x_4	x_5	x_6	x_7	rhs
$-\frac{7}{12}$	1	0	$\frac{1}{3}$	$-\frac{1}{4}$	0	0	$\frac{11}{12}$
$\frac{11}{6}$	0	1	$-\frac{1}{3}$	$\frac{1}{2}$	0	0	$\frac{5}{6}$
$\frac{19}{4}$	0	0	0	$-\frac{1}{4}$	1	0	$\frac{41}{4}$
$\frac{7}{12}$	0	0	$-\frac{1}{3}$	$\frac{1}{4}$	0	1	$-\frac{11}{12}$
$\frac{17}{12}$	0	0	$\frac{1}{3}$	$\frac{3}{4}$	0	0	$\frac{95}{12}$

Table 2.4

x_1	x_2	x_3	x_4	x_5	x_6	x_7	rhs
0	1	0	0	0	0	0	0
$\frac{5}{4}$	0	1	0	$\frac{1}{4}$	0	0	$\frac{7}{4}$
$\frac{19}{4}$	0	0	0	$-\frac{1}{4}$	1	0	$\frac{41}{4}$
$-\frac{7}{4}$	0	0	1	$-\frac{3}{4}$	0	1	$\frac{11}{4}$
2	0	0	0	1	0	0	7

obtain:

$$x_1^* = \tfrac{1}{7},$$
$$x_2^* = 1,$$
$$x_3^* = \tfrac{4}{7},$$
$$Z^* = \tfrac{54}{7}.$$

This also is infeasible for the IP so let us review the situation. From the solution to (I) we know that the best *feasible* IP solution with $x_2 = 0$ cannot have a value better than 7 (the value of the LP solution). From the solution to (II) we know that the best *feasible* IP solution with $x_2 \geq 1$ cannot have a value better than $\frac{54}{7}$ (the value of the LP solution). Consider the set of all the feasible solutions to the original IP, say S. We can partition S into two disjoint subsets, S_1 and S_2, where all solutions in S_1 have $x_2 = 0$ and all solutions in S_2 have $x_2 \geq 1$. Then the value of the best solution in $S_1(S_2)$ cannot be more than 7 ($\frac{54}{7}$). Naturally the value of the best solution may turn out to be strictly less than these numbers, but we have established an *upper bound* on the values of the solutions in each set.

Because the upper bound of S_2 is higher than that of S_1, S_2 will be further examined. In order to keep track of the information that is uncovered, a *decision tree* is used, as displayed in Fig. 2.2.

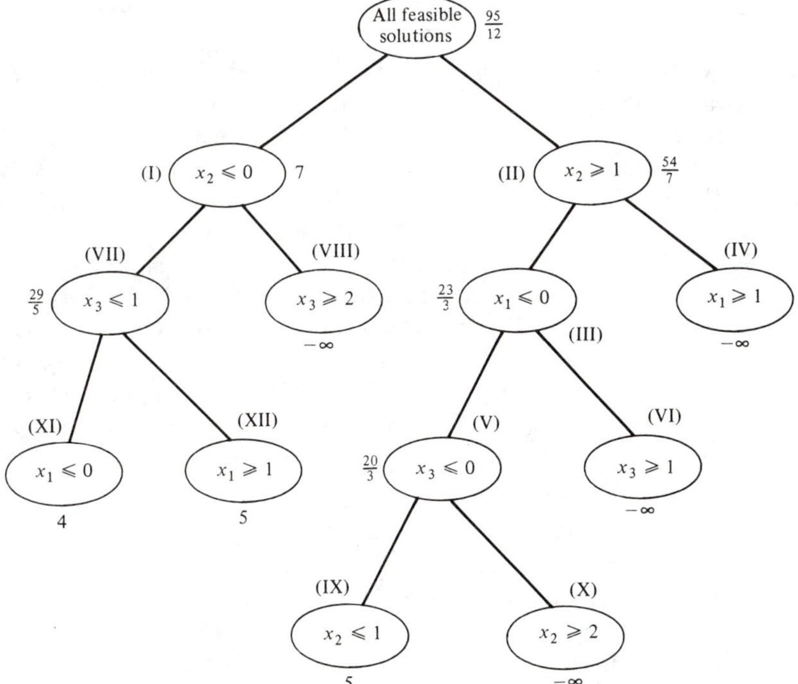

Figure 2.2. A Decision Tree for Dakin's Method.

Branching from node (II) we choose a variable with a noninteger value, say $x_1^* = \frac{1}{7}$. Then proceeding as before we create two new constraints:

$$x_1 \leq 0$$

and

$$x_1 \geq 1.$$

These are then added separately to the LP problem (II) to create two new problems: (III) and (IV). These new problems can be solved most easily by applying the dual simplex method to the final tableau of problem (II). As can be seen, the bound on the best IP value (i.e., the LP optimal-solution value) is associated with each node. As problem (IV) has no feasible solutions, the symbol $-\infty$ is used to designate its upper bound.

Branching from node (III) [as its bound of $\frac{23}{3}$ is higher than that of node (I)] we create nodes (V) and (VI). The highest bound is now that of node (I); we branch from it to create nodes (VII) and (VIII). The highest bound is now that of node (V) and so we create nodes (IX) and (X). The LP solution for node (IX) is

$$x_1^* = 0,$$

$$x_2^* = 1,$$

$$x_3^* = 0,$$

$$x_5^* = 5,$$

which is feasible for the original IP. This is the best IP solution we have found so far, indeed it is the only one. It is stored as the *incumbent* and so we proceed, branching from node (VII). [There is no reason to branch from node (IX) as a feasible solution has been found for it.] This creates nodes (XI) and (XII) with bounds no better than the value of the incumbent. Thus there is no reason to branch further from these nodes as no solution better than the incumbent could be unearthed. Thus we have shown that the incumbent must be optimal, so the search is terminated.

We have found that the optimal solution to this problem is to fly one B crate, no others, for a revenue of five units. Let us now review some of the general principles of branch-and-bound enumeration. During the course of the search, a node that is currently at the endpoint of a path from the node labeled *afs* (all feasible solutions) must be in one of the following states, where its associated IP problem:

(i) Has no feasible solution.
(ii) Is such that its LP solution S is feasible and
  (a) S has a value better than that of the incumbent, or
  (b) S has a value no better than that of the incumbent.
(iii) Neither of states (i) or (ii) have as yet been established.

So that state (ii) can be attained before an initial incumbent has been found, define

$$Z = \text{the value of the current incumbent solution,}$$

and initially set

$$Z = -\infty.$$

A node in either state (i) or (ii) is said to be *fathomed*. A node that has not yet been fathomed is said to be *active*. A node in states (i) or (iib) represents a set of feasible IP solutions, which cannot possibly contain the optimum. It is thus disregarded, or "pruned" from the tree. The search continues, branching from the active node with the highest bound until either:

(i) All nodes have been pruned from the tree (in which case the problem has no feasible solution) or
(ii) All active nodes have bounds no greater than the value of the incumbent (in which case the incumbent represents an optimal solution).

A summary of Dakin's method is given in the flowchart of Fig. 2.3.

2.1.2.2. *The Method of Balas*. A second branch-and-bound method will also be discussed. It is the method of Balas (1965) which is designed for zero–one IP problems. In order to motivate it we return to the airfreight problem.

Suppose now the management of the Quick as Lightning company decides to make some major policy changes regarding their problem. Their old DC3 is sold and a new, larger aircraft is bought which has volume and weight capacity of 8 and 10 units, respectively. The constraint concerning insurance is unchanged. However, it has now decreed that at most one of each type of crate will be carried per flight. The problem becomes

$$\text{Maximize } 3x_1 + 5x_2 + 4x_3 \quad (= Z) \tag{2.12}$$

subject to

$$2x_1 + 6x_2 + 3x_3 \le 8, \tag{2.13}$$

$$5x_1 + 4x_2 + 4x_3 \le 10, \tag{2.14}$$

$$6x_1 + x_2 + x_3 \le 12, \tag{2.15}$$

$$x_1, x_2, x_3 = 0 \quad \text{or} \quad 1. \tag{2.16}$$

This is a zero–one IP problem and we now use the method of Balas to solve it. As the method is of the branch-and-bound type, a decision tree will be built up. The reader should refer to Fig. 2.4 during this discussion. We begin by assigning a value of one to all variables. That is, set

$$(x_1, x_2, x_3) = (1,1,1). \tag{2.17}$$

Because all the objective-function coefficients are nonnegative, if (2.17) was

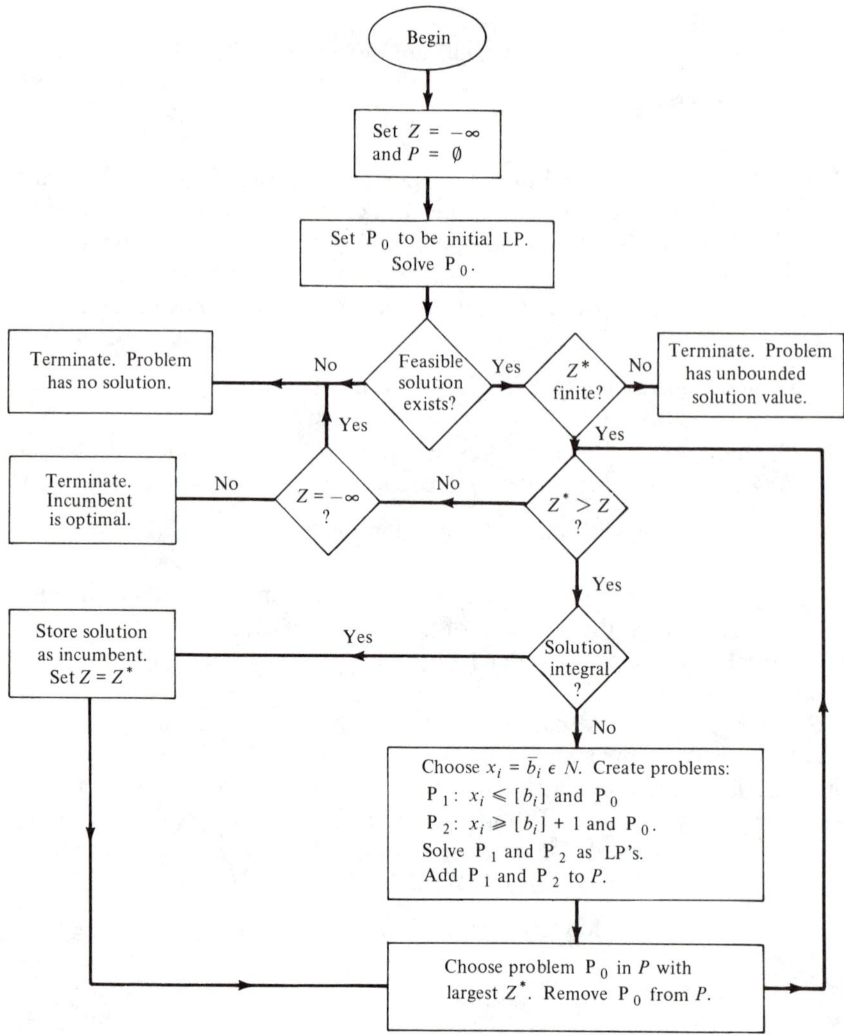

Figure 2.3. Summary of Dakin's Method.

feasible it would be optimal. Indeed Balas's method assumes

$$c_i \geq 0, \qquad i = 1, 2, \ldots, n.$$

Any IP problem can be transformed so that this is so without altering its set of feasible solutions. This point will be raised in the general discussions later.

However, (2.17) is infeasible so at least one variable must be zero in any feasible solution. The variable whose reduction to zero will reduce Z by the least will be the one with the smallest objective-function coefficient. This is x_1 with

$$c_1 = 3.$$

Thus the largest possible feasible-solution value we can hope for is the sum of the other coefficients, $4+5=9$. We have established an upper bound on the value of the optimal solution, which is associated with the initial node (*afs*) in the decision tree in Fig. 2.4. We now make a decision about x_1. Like all variables, it has value either 0 or 1. We can partition the set of feasible solutions into two sets—one set in which

$$x_1 = 0$$

and the other in which

$$x_1 = 1.$$

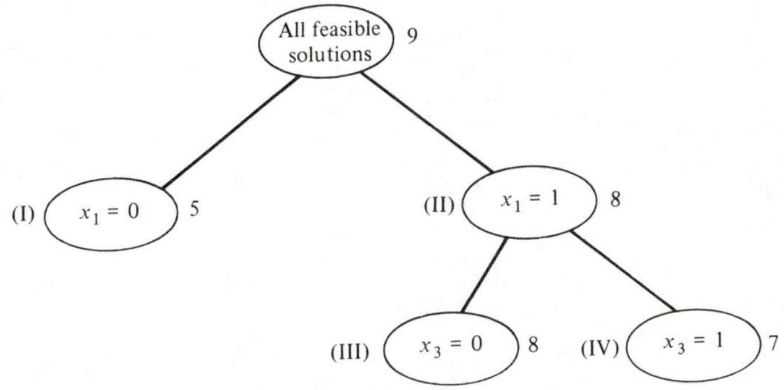

Figure 2.4. A Decision Tree for the Method of Balas.

These sets are represented by nodes (I) and (II) in the tree.

Let us consider (I) first. Given

$$x_1 = 0,$$

we set all unspecified variables equal to one and check to see whether the resulting solution is feasible. Now

$$(x_1, x_2, x_3) = (0, 1, 1)$$

is infeasible as (2.13) is violated. Hence the largest feasible solution we can hope for with

$$x_1 = 0$$

is obtained by ignoring the smallest objective-function coefficient among those belonging to unspecified variables, namely, x_3 with

$$c_3 = 4.$$

The bound for (I) is thus 5. Examining (II), we know that (2.17) is infeasible so we set the unspecified variable with least c_j to zero. Thus set

$$x_3 = 0$$

and obtain a bound of

$$3 + 5 = 8,$$

that is, the sum of: the coefficients of all variables set equal to one and the coefficients of all unspecified variables.

Adopting a policy of branching from the node with the highest bound we branch from (II). The unspecified variable with least c_j is x_3. We create two new nodes, (III) and (IV), where

$$x_3 = 0 \quad (\text{and } x_1 = 1)$$

and

$$x_3 = 1 \quad (\text{and } x_1 = 1),$$

for (III) and (IV), respectively. Examining (III) first we set the unspecified variables (x_2 only) to one and discover that

$$(x_1, x_2, x_3) = (1,1,0) \tag{2.18}$$

is feasible with a value of 8. This solution is stored as the incumbent. In (IV) we know (2.17) is infeasible thus x_2 is set to zero producing a bound of 7. Node (III) represents the optimal solution as the value of its feasible solution is no less than the bound of any other node in the tree. The optimal solution is

$$x_1^* = 1,$$

$$x_2^* = 1,$$

$$x_3^* = 0,$$

$$Z^* = 8.$$

In terms of our original problem, it appears that the company should fly one A and one B crate on each trip for a return of eight units.

The method of Balas is additive in the sense that it requires only the addition (not multiplication) of numbers in the data. It assumes that all objective function coefficients are nonnegative. If this assumption is violated and there exists an i, $1 \le i \le n$, such that

$$c_i < 0,$$

then x_i can be replaced throughout the formulation by $\bar{x}_i$ where

$$\bar{x}_i = 1 - x_i.$$

For each node in the decision tree, the variables are partitioned into three mutually exclusive subsets:

$$V = \text{the set of variables set equal to zero;}$$

$$W = \text{the set of variables set equal to one;}$$

$$F = \text{the free variables, not yet specified.}$$

Initially, set

$$F = \{x_1, x_2, \dots, x_n\},$$

$$V = W = \varnothing.$$

In general, when examining a node representing an assignment to V, W, and F, all variables in F are temporarily assigned a value of 1. If this solution is feasible, it is clearly the best that can be obtained for the particular partition of V, W, and F, as

$$c_i \geq 0, \qquad i = 1, 2, \ldots, n.$$

If this solution is infeasible, an upper bound can be obtained for the set of values for all feasible solutions with the policy of V and W. This bound is

$$\sum_{x_i \in F} c_i - \operatorname*{Min}_{x_i \in F} \{c_i\} + \sum_{x_i \in W} c_i.$$

Two new nodes are sprouted from a node with bound greater than that of the incumbent solution value. For one node, the sets V, W, and F remain as they are in the parent node except that

$$F \text{ becomes } F \backslash \{x_j\}$$

and

$$W \text{ becomes } W \cup \{x_j\}.$$

And in the other node,

$$F \text{ becomes } F \backslash \{x_j\}$$

and

$$V \text{ becomes } V \cup \{x_j\}.$$

where x_j is such that $c_j = \operatorname{Min}_{x_i \in F}\{c_i\}$.

A flowchart for the method of Balas is given in Fig. 2.5.

2.1.3. Cutting Planes

In this section we develop an alternative way to solve IP problems. It is introduced via the airfreight problem: (2.1)–(2.5). Suppose that (2.1)–(2.4) is solved by the simplex method. The final tableau was given in Table 2.1. It can be seen that x_2 and x_3 are noninteger. Let us examine the top line in Table 2.1, corresponding to x_2, which must be satisfied by any feasible solution:

$$-\tfrac{7}{12}x_1 + x_2 + \tfrac{1}{3}x_4 - \tfrac{1}{4}x_5 = \tfrac{11}{12}. \tag{2.19}$$

Equation (2.19) can be solved for x_2:

$$x_2 = \tfrac{11}{12} - \left(-\tfrac{7}{12}\right)x_1 - \tfrac{1}{3}x_4 - \left(-\tfrac{1}{4}\right)x_5. \tag{2.20}$$

Each number on the right-hand side of the equation can be expressed as a sum of an integer and a positive fraction:

$$x_2 = 0 + \tfrac{11}{12} - \left(-1 + \tfrac{5}{12}\right)x_1 - \left(0 + \tfrac{1}{3}\right)x_4 - \left(-1 + \tfrac{3}{4}\right)x_5. \tag{2.21}$$

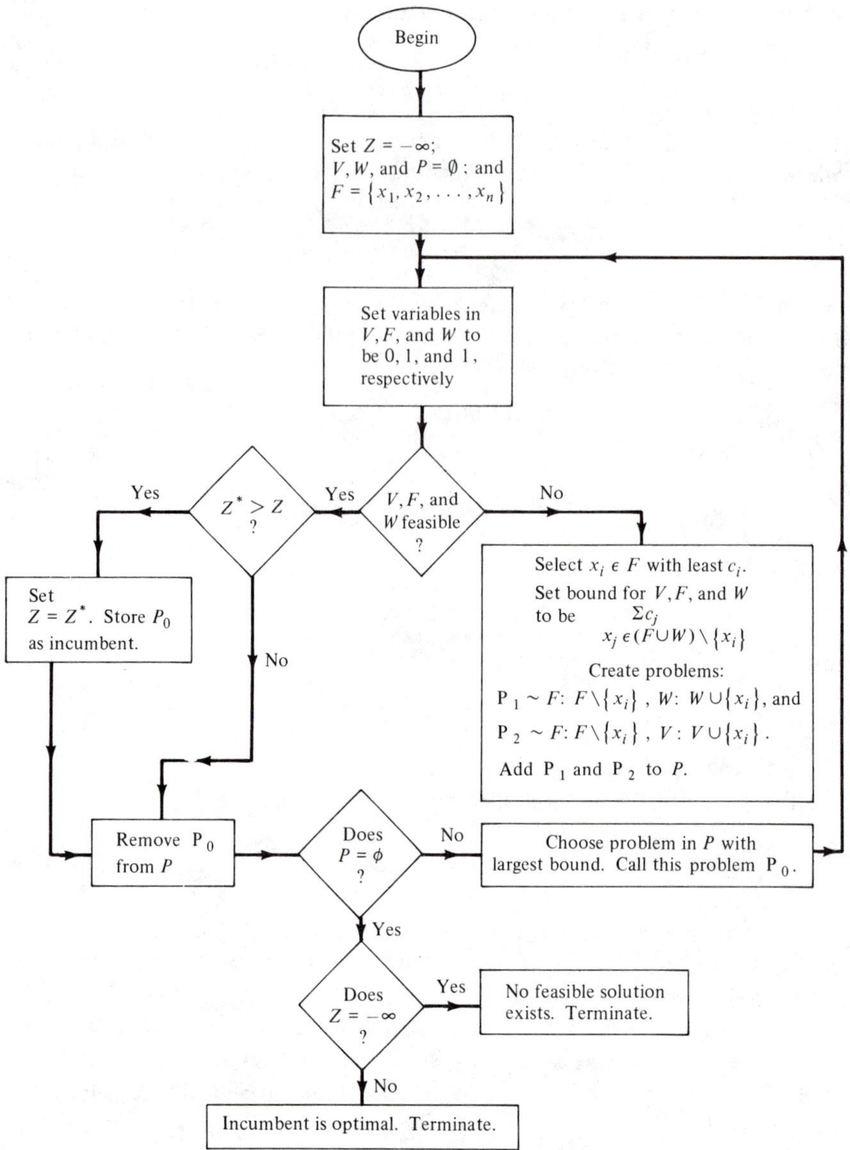

Figure 2.5. A Flowchart for the Method of Balas.

The integer parts and the fractional parts of (2.21) can be collected together:

$$x_2 = \{0-(-1)x_1-(0)x_4-(-1)x_5\} + \{\tfrac{11}{12} - \tfrac{5}{12}x_1 - \tfrac{1}{3}x_4 - \tfrac{3}{4}x_5\}.$$

Because (2.2), (2.3), and (2.4) have integer coefficients, the slack variables x_3, x_4, and x_5 must be integer in any feasible solution. Thus the expression:

$$\{0-1(-1)x_1-(0)x_4-(-1)x_5\}$$

must be integral in any feasible solution. As x_2 must be an integer, the second expression:

$$\left\{\tfrac{11}{12} - \tfrac{5}{12}x_1 - \tfrac{1}{3}x_4 - \tfrac{3}{4}x_5\right\}$$

must be an integer in any feasible solution. However, it is the difference between a fraction ($\frac{11}{12}$) and a sum of fractions, times variables constrained to be integer. The largest value this expression can attain is $\frac{11}{12}$ when

$$x_1 = x_4 = x_5 = 0.$$

However, as it is constrained to be integral it must be a nonpositive integer. Therefore

$$\tfrac{11}{12} - \tfrac{5}{12}x_1 - \tfrac{1}{3}x_4 - \tfrac{3}{4}x_5 \leq 0. \tag{2.22}$$

Equation (2.22) is a further constraint that must be satisfied by any feasible solution. We can add a slack variable x_7 to it:

$$-\tfrac{5}{12}x_1 - \tfrac{1}{3}x_4 - \tfrac{3}{4}x_5 + x_7 = -\tfrac{11}{12}$$

and add it to Table 2.1 to produce Table 2.5. Applying the dual simplex method, Table 2.6 is produced. This process can now be repeated with the equation for x_3:

$$\tfrac{9}{4}x_1 + x_3 + \tfrac{5}{4}x_5 - x_7 = \tfrac{7}{4}.$$

Solving for x_3:

$$x_3 = (1+\tfrac{3}{4}) - (2+\tfrac{1}{4})x_1 - (1+\tfrac{1}{4})x_5 - (-1+0)x_7.$$

Table 2.5

x_1	x_2	x_3	x_4	x_5	x_6	x_7	rhs
$-\frac{7}{12}$	1	0	$\frac{1}{3}$	$-\frac{1}{4}$	0	0	$\frac{11}{12}$
$\frac{11}{6}$	0	1	$-\frac{1}{3}$	$\frac{1}{2}$	0	0	$\frac{5}{6}$
$\frac{19}{4}$	0	0	0	$-\frac{1}{4}$	1	0	$\frac{41}{4}$
$-\frac{5}{12}$	0	0	$-\frac{1}{3}$	$-\frac{3}{4}$	0	1	$-\frac{11}{12}$
$\frac{17}{12}$	0	0	$\frac{1}{3}$	$\frac{3}{4}$	0	0	$\frac{95}{12}$

Table 2.6

x_1	x_2	x_3	x_4	x_5	x_6	x_7	rhs
-1	1	0	0	-1	0	1	0
$\frac{9}{4}$	0	1	0	$\frac{5}{4}$	0	-1	$\frac{7}{4}$
$\frac{19}{4}$	0	0	0	$-\frac{1}{4}$	1	0	$\frac{41}{4}$
$\frac{5}{4}$	0	0	1	$\frac{9}{4}$	0	-3	$\frac{11}{4}$
1	0	0	0	0	0	1	7

The same reasoning as above produces the constraint:

$$\tfrac{3}{4} - \tfrac{1}{4}x_1 - \tfrac{1}{4}x_5 \leq 0. \tag{2.23}$$

On introduction of the slack variable x_8, this becomes

$$-\tfrac{1}{4}x_1 - \tfrac{1}{4}x_5 + x_8 = -\tfrac{3}{4}.$$

Introducing this into Table 2.6 we arrive at Table 2.7. We now apply the dual simplex method to Table 2.7 to get Tables 2.8–2.10. Table 2.10 represents a feasible solution

$$x_1^* = 0, \qquad x_2^* = 1, \qquad x_3^* = 0, \qquad Z^* = 5.$$

As only essential constraints have been added to the original problem, this

Table 2.7

x_1	x_2	x_3	x_4	x_5	x_6	x_7	x_8	rhs
-1	1	0	0	-1	0	1	0	0
$\frac{9}{4}$	0	1	0	$\frac{5}{4}$	0	-1	0	$\frac{7}{4}$
$\frac{19}{4}$	0	0	0	$-\frac{1}{4}$	1	0	0	$\frac{41}{4}$
$\frac{5}{4}$	0	0	1	$\frac{9}{4}$	0	-3	0	$\frac{11}{4}$
$-\frac{1}{4}$	0	0	0	$-\frac{1}{4}$	0	0	1	$-\frac{3}{4}$
1	0	0	0	0	0	1	0	7

Table 2.8

x_1	x_2	x_3	x_4	x_5	x_6	x_7	x_8	rhs
0	1	0	0	0	0	1	-4	3
1	0	1	0	0	0	-1	5	-2
5	0	0	0	0	1	0	-1	11
-1	0	0	1	0	0	-3	9	-4
1	0	0	0	1	0	0	-4	3
1	0	0	0	0	0	1	0	7

Table 2.9

x_1	x_2	x_3	x_4	x_5	x_6	x_7	x_8	rhs
$-\frac{1}{3}$	1	0	$\frac{1}{3}$	0	0	0	-1	$\frac{5}{3}$
$\frac{4}{3}$	0	1	$-\frac{1}{3}$	0	0	0	2	$-\frac{2}{3}$
5	0	0	0	0	1	0	-1	11
$\frac{1}{3}$	0	0	$-\frac{1}{3}$	0	0	1	-3	$\frac{4}{3}$
1	0	0	0	1	0	0	-4	3
$\frac{2}{3}$	0	0	$\frac{1}{3}$	0	0	0	3	$\frac{17}{3}$

Table 2.10

x_1	x_2	x_3	x_4	x_5	x_6	x_7	x_8	rhs
1	1	1	0	0	0	0	1	1
−4	0	−3	1	0	0	0	−6	2
5	0	0	0	0	1	0	−1	11
−1	0	−1	0	0	0	1	−5	2
1	0	0	0	1	0	0	−4	3
2	0	1	0	0	0	0	5	5

solution must be optimal. It is, of course, the same solution as that obtained by Dakin's method.

Let us now develop the general theory of the cutting-plane approach as introduced by Gomory (1958). The method assumes that all the data are rational. The implications of this are that fractions can be cleared so that all constraints have integer coefficients. New constraints called *cuts* are introduced into the problem one at a time which progressively remove noninteger parts of the feasible region. No feasible integer solution is ever removed by a cut. Examples of these cuts are (2.22) and (2.23).

The method begins by solving the underlying LP problem by the simplex method. As in Dakin's method, if the LP solution is integer, it is optimal and nothing further needs to be done. If it is noninteger, the equation corresponding to a noninteger variable is identified, say the jth row:

$$x_i + \bar{a}_{j_1} y_1 + \bar{a}_{j2} y_2 + \cdots + \bar{a}_{j_n} y_n = \bar{b}_j, \tag{2.24}$$

where x_i is a basic variable and $y_1, y_2, \ldots, y_n$ represent the nonbasic x_i's. Equation (2.24) can be rewritten as

$$x_i = \bar{b}_j - \sum_{k=1}^{n} \bar{a}_{j_k} y_k, \tag{2.25}$$

and each of the numbers $\bar{b}_j$ and $\bar{a}_{j_1}, \bar{a}_{j_2}, \ldots, \bar{a}_{j_n}$ can be expressed as the sum of the greatest integer not exceeding it and a fractional part.

Let the integral part of a number h be denoted by $[h']$ and the fractional part by Z'. Then (2.25) becomes

$$x_i = [\bar{b}_j] + \bar{b}'_j - \sum_{k=1}^{n} \left([\bar{a}_{jk}] + \bar{a}'_{jk}\right) y_k.$$

This can be written as

$$\begin{aligned} x_i &= \left\{[\bar{b}_j] - \sum_{k=1}^{n} [\bar{a}_{jk}] y_k\right\} + \left\{\bar{b}'_j - \sum_{k=1}^{n} \bar{a}'_{jk} y_k\right\} \\ &= \alpha + \beta, \end{aligned}$$

where

$$\alpha = [\bar{b}_j] - \sum_{k=1}^{n} [\bar{a}_{jk}] y_k$$

and

$$\beta = \bar{b}_j' - \sum_{k=1}^{n} \bar{a}_{jk}' y_k. \tag{2.26}$$

As integral coefficients have been assumed, all y_k will be integer in any feasible solution. Thus α will be integer for any feasible solution. As x_i must be integer in any feasible solution, β must also be integer. Let us examine (2.26). Because $\bar{b}_j$ is noninteger,

$$0 < \bar{b}_j' < 1.$$

Also

$$0 \leq \bar{a}_{jk}', \qquad k = 1, 2, \ldots, n,$$

and

$$0 \leq y_k, \qquad k = 1, 2, \ldots, n.$$

Hence β attains its maximum value when

$$y_k = 0, \qquad k = 1, 2, \ldots, n.$$

$$\therefore \quad \beta \leq \bar{b}_j'.$$

But as $\bar{b}_j'$ is fractional and β must be integer,

$$\beta \leq 0.$$

That is,

$$\bar{b}_j' - \sum_{k=1}^{n} a_{jk}' y_k \leq 0. \tag{2.27}$$

Equation (2.27) is termed a Gomory cut and must be satisfied by any feasible solution. It is added to the original problem as follows. A slack variable x_q is introduced into (2.27):

$$\sum_{k=1}^{n} \bar{a}_{jk}' y_k + x_1 = -\bar{b}_j'.$$

This is then added to the final simplex tableau and the dual simplex method is used to produce another final tableau. If this is feasible, it is optimal and the procedure is terminated. Otherwise, another noninteger variable is identified and the process is repeated.

Consider now a mixed-integer programming problem. That is, some but not all of the variables are constrained to be integers. In terms of (2.6)–(2.9),

$$0 < q < n.$$

Gomory's mixed-integer LP algorithm follows the same initial pattern as the all-integer algorithm. Suppose the initial simplex solution contains a noninteger-valued variable x_j, which is one of those which is constrained to be integers. Then its tableau equation can be rewritten as

$$\left[\bar{b}_j\right] + \bar{b}_j' - x_j = \sum_{k=1}^{p} \bar{a}_{jk} y_k. \tag{2.28}$$

At this point the analysis takes a different path from the all-integer case because not all of the variables y_k, $k=1,\ldots,p$, may be constrained to be integers. However let

$$S_+ = \{k : \bar{a}_{jk} \geq 0\}$$

and

$$S_- = \{k : \bar{a}_{jk} < 0\}.$$

Then (2.28) can be written as

$$\left[\bar{b}_j\right] + \bar{b}_j' - x_j = \sum_{k \in S_+} \bar{a}_{jk} y_k + \sum_{k \in S_-} \bar{a}_{jk} y_k. \tag{2.29}$$

Case I

Assume $[\bar{b}_j] + \bar{b}_j' - x_j < 0$. Now as $[\bar{b}_j]$ is an integer, x_j is constrained to be an integer in any feasible solution, and $\bar{b}_j'$ is a nonnegative fraction. Thus

$$\left[\bar{b}_j\right] - x_j$$

must be a negative integer, say $-u$.

$$\therefore \left[\bar{b}_j\right] + \bar{b}_j' - x_j = \bar{b}_j' - u, \qquad u \in \{1,2,3,\ldots\}.$$

Using (2.29),

$$\bar{b}_j' - u = \sum_{k \in S_+} \bar{a}_{jk} y_k + \sum_{k \in S_-} \bar{a}_{jk} y_k.$$

Now since

$$u \geq 1,$$

we have

$$\bar{b}_j' - 1 \geq \sum_{k \in S_+} \bar{a}_{jk} y_k + \sum_{k \in S_-} \bar{a}_{jk} y_k.$$

And from the definition of S_+ and the fact that $y_k \geq 0$ for all k,

$$\bar{b}_j' - 1 \geq \sum_{k \in S} \bar{a}_{jk} y_k.$$

Now as $\bar{b}_j' - 1 < 0$,

$$1 \leq \left(\bar{b}_j' - 1\right)^{-1} \sum_{k \in S_-} \bar{a}_{jk} y_k,$$

and multiplying by $\bar{b}_j'$,

$$\bar{b}_j' \leq \bar{b}_j' \left(\bar{b}_j' - 1\right)^{-1} \sum_{k \in S_-} \bar{a}_{jk} y_k. \tag{2.30}$$

Case II

Assume $[\bar{b}_j] + \bar{b}_j' - x_j \geq 0$. As x_j is constrained to be an integer in any feasible solution

$$\left[\bar{b}_j\right] + \bar{b}_j' - x_j = \bar{b}_j' + v$$

for some v, where $v \in \{0, 1, 2, 3, \ldots\}$.

Using (2.29),

$$\bar{b}_j' + v = \sum_{k \in S_+} \bar{a}_{jk} y_k + \sum_{k \in S_-} \bar{a}_{jk} y_k.$$

Now since $v \geq 0$, we have

$$\bar{b}_j' \leq \sum_{k \in S_+} \bar{a}_{jk} y_k + \sum_{k \in S_-} \bar{a}_{jk} y_k.$$

And from the definition of S_- and the fact that $y_k \geq 0$ for all k,

$$\bar{b}_j' \leq \sum_{k \in S_+} \bar{a}_{jk} y_k. \tag{2.31}$$

Combining (2.30) and (2.31),

$$\bar{b}_j' \leq \bar{b}_j' \left(\bar{b}_j' - 1\right)^{-1} \sum_{k \in S_-} \bar{a}_{jk} y_k + \sum_{k \in S_+} \bar{a}_{jk} y_k. \tag{2.32}$$

This inequality must be satisfied if x_i is to be an integer. The constraint (2.32) is the Gomory cut, which is now introduced into the final tableau.

A slack variable x_r is now added to (2.32):

$$\bar{b}_j' = \bar{b}_j' \left(\bar{b}_j' - 1\right)^{-1} \sum_{k \in S_-} \bar{a}_{jk} y_k + \sum_{k \in S_+} \bar{a}_{jk} y_k - x_r. \tag{2.33}$$

Now as

$$y_k = 0, \qquad k = 1, 2, \ldots, p,$$

we have

$$x_r = -\bar{b}_j',$$

which is infeasible. The dual simplex method is used to remedy this situation. The above process is repeated until either one of the following occurs.

(i) A tableau is produced in which x_i, $i=1,2,\ldots,q$, are integer in which case the corresponding solution is optimal.
(ii) The use of the dual simplex method leads to the conclusion that no feasible solution exists in which case one can conclude that the original mixed-integer problem has no feasible solution.

The method will be illustrated on the following problem.

$$\text{Maximize } 4x_1+3x_2 \quad (=Z) \tag{2.34}$$

subject to

$$3x_1+4x_2\le 12,$$
$$4x_1+2x_2\le 9, \tag{2.35}$$
$$x_1, x_2 \ge 0, \tag{2.36}$$
$$x_1 \text{ must be an integer.} \tag{2.37}$$

That is,

$$q=1.$$

On examining the optimal simplex tableau (Table 2.11) for (2.34)–(2.36) it can be seen that x_1 is noninteger and can be expressed as

$$1+\tfrac{1}{5}-x_1=-\tfrac{1}{5}x_3+\tfrac{2}{5}x_4.$$

Table 2.11

x_1	x_2	x_3	x_4	b
0	1	$\frac{2}{5}$	$-\frac{3}{10}$	$\frac{21}{10}$
1	0	$-\frac{1}{5}$	$\frac{2}{5}$	$\frac{6}{5}$
0	0	$\frac{2}{5}$	$\frac{7}{10}$	$\frac{11}{10}$

In terms of (2.28):

$$[\bar{b}_j]=1,$$
$$\bar{b}_j'=\tfrac{1}{5},$$
$$j=2,$$
$$i=1,$$
$$p=2,$$
$$\bar{a}_{j_1}=-\tfrac{1}{5},$$
$$\bar{a}_{j_2}=\tfrac{2}{3},$$
$$y_1=x_3,$$
$$y_2=x_4.$$

Also

$$S_+ = \{4\}$$

and

$$S_- = \{3\}.$$

Letting

$$x_r = x_5,$$

in terms of (2.33) the cut becomes

$$\tfrac{1}{5} = \tfrac{1}{5}(\tfrac{1}{5}-1)^{-1}(-\tfrac{1}{5})x_3 + \tfrac{2}{5}x_4 - x_5.$$

Adding the negative of this constraint to Table 2.11 yields Table 2.12. The application of the dual simplex method yields Table 2.13. Table 2.13 displays the optimal solution to the problem as x_1 is now integer-valued. This solution is

$$x_1^* = 1,$$

$$x_2^* = \tfrac{9}{4},$$

$$Z^* = \tfrac{43}{4}.$$

Table 2.12

x_1	x_2	x_3	x_4	x_5	b
0	1	$\frac{2}{5}$	$-\frac{3}{10}$	0	$\frac{21}{10}$
1	0	$-\frac{1}{5}$	$\frac{2}{5}$	0	$\frac{6}{5}$
0	0	$\frac{1}{20}$	$-\frac{2}{5}$	1	$-\frac{1}{5}$
0	0	$\frac{2}{5}$	$\frac{7}{10}$	0	$\frac{111}{10}$

Table 2.13

x_1	x_2	x_3	x_4	x_5	b
0	1	$\frac{47}{80}$	0	$-\frac{3}{4}$	$\frac{9}{4}$
1	0	$-\frac{9}{20}$	0	1	1
0	0	$\frac{1}{8}$	1	$-\frac{5}{2}$	$\frac{1}{2}$
0	0	$\frac{67}{80}$	0	$\frac{7}{4}$	$\frac{43}{4}$

2.1.4. Exercises

1. Solve the following problem by Dakin's method:

$$\text{Maximize } 4x_1 + 3x_2 + 3x_3$$

subject to

$$\begin{aligned}4x_1+2x_2+\ x_3&\le 10,\\3x_1+4x_2+2x_3&\le 14,\\2x_1+\ x_2+3x_3&\le 7,\end{aligned}$$

x_1, x_2, x_3, nonnegative integers.

2. Solve problem 1 above by the method of Balas by converting to zero–one variables.
3. Add the following constraint to problem 1 above:

$$x_1, x_2, x_3 = 0 \quad \text{or} \quad 1.$$

   Solve the resulting problem by the method of Balas.
4. Solve problem 1 above by Gomory's method.
5. Suppose in problem 1 above that only x_1 is constrained to be integral. Solve this new problem by Gomory's mixed-integer method.
6. Solve problem 3 above by exhaustive enumeration. Compare the amount of computation required with that required by the method of Balas.

*7. Formulate the traveling-salesman-path problem of Chapter 0 as an integer-programming problem.

2.2. Dynamic Programming

2.1.1. A Simple Dynamic-Programming Problem

Consider Fig. 2.6, ignoring the boxed numbers, which depicts a network of huts connected by trails. Each path is labeled with its distance in kilometers. Suppose we have a hiker who would like to trudge from hut 1 to hut 11 by the shortest path of trails. The hiker could, of course, list all possible paths from 1 to 11, calculate their distances, and choose the shortest. But such a method is useful only for relatively small networks. If the number of huts in the problem is gradually increased, the amount of computation increases very quickly indeed. Soon it is far too much for hand calculation and after that it is inefficient even for a computer. Further discussion on the inefficiency of the fundamental algorithm appears in Section 2.3.

Obviously we must find a method with less computation if we are to be able to guarantee to find shortest paths for large networks. The technique of *dynamic programming* (DP) can be used to devise such a method. Other shortest-path methods are given in Chapter 3. Dynamic programming is a generalized approach to solving staged optimization and can be applied to a tremendous variety of problems, not just to finding shortest paths. We now apply it to the problem of Fig. 2.6.

We begin by dividing the huts into *stages*, as shown in Fig. 2.6. The hiker will visit exactly one hut from each stage on the trip. Each stage is numbered to represent the number of trails walked on the trip so far to get to that stage. For each stage we define at least one *state*. The states for each stage are the numbers of the huts associated with that stage. For example, the states of stage 1 are 2, 3, and 4. Naturally the hiker will be in exactly one state (the number of the hut the hiker visits) at each stage. We define for each state a *return*. The return of a state is the minimum distance from hut

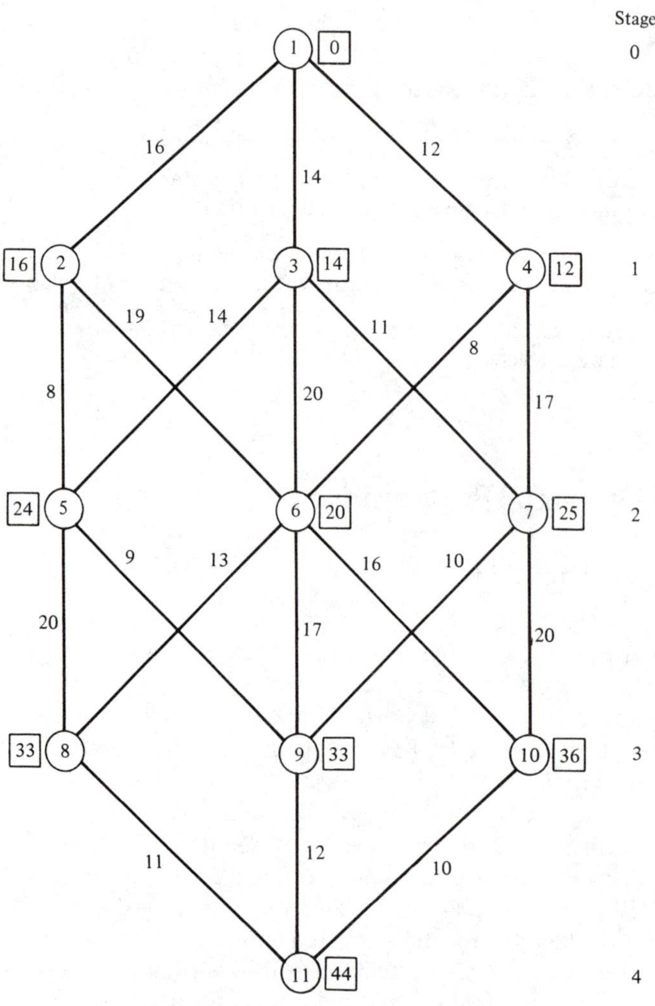

Figure 2.6. The Network of Trails for the Hiker's Problem.

1 to the hut of that state. Let us now solve the problem by using these ideas. The reader should refer to Fig. 2.6 during the course of the discussion where the various states are depicted in boxes.

Initially, the hiker is at stage 0 in state 1. The hiker walks a first trail and is at stage 1 in one of the states 2, 3, or 4. The returns for states 2, 3, and 4 are 16, 14, and 12, respectively, the distances of these first trails. The hiker now proceeds from stage 1 to stage 2. At stage 2 the hiker will be in either of the states 5, 6, or 7. Let us suppose for the moment that the hiker ends up in state 5. There are two ways to get there: from state 2 or state 3. If the hiker comes from state 2 the total distance walked will be the sum of the state 2 return and the length of 2–5 trail, that is, $16+8=24$. If one arrives via state 3 the total distance traveled will be $14+14=28$.

The following idea is an example of the fundamental concept of dynamic programming. Once the hiker leaves state 5, the later trail choices are not affected by the actual path which was taken to get to 5. One just wants to get from 5 to 11 by the shortest route. To minimize overall distance the hiker should proceed from 1 to 5 by the shortest 1–5 path. Any shortest 1–11 path via 5 will contain the shortest 1–5 path, otherwise it could be improved. So we can look at the two options for getting to 5 and choose the shorter, namely, via 2 for a distance of 24 km. By definition this becomes the return for state 5.

We now relax our assumption that the hiker was in state 5. Similar reasoning can be used to deduce that the states 6 and 7 returns are the minimum of $(16+19, 14+20, 12+8)$ and the minimum of $(14+11, 12+17)$, respectively. The hiker now goes from stage 2 to 3. The stage 3 returns are calculated in the same manner as before. For example, the state 8 return is the minimum of the state 5 return plus the 5–8 distance and the state 6 return plus the 6–8 distance. In order to calculate the state 8 return we need to know only the states 5 and 6 return, not the paths that produced them. The reader should verify that the returns for states 8, 9, and 10 are 33, 33, and 36 km, respectively.

The hiker now comes to the final stage. Its only state (11) has a return of 44. We now know that the length of the shortest path is 44 km. To find the path itself we must backtrack, using the returns and trail distances. Where did the state 11 return of 44 come from? We can test to see if the state 8 return plus the 8–11 trail length equals the state 11 return. If it does, hut 8 is on a shortest path. If it does not, hut 8 is not on a shortest path. We can test each of the other stage 3 returns in a similar manner. Any returns for which the above test is positive have huts which are on a shortest path. (There may be more than one shortest path.) In our present case the test is positive only for state 8. So we know that the shortest path goes through hut 8 just before reaching hut 11. We now ask where the state 8 return came from. Working back in this way we find that the complete shortest path is $1, 4, 6, 8, 11$. In solving this problem we have used all the notions of basic DP.

2.2.2. Forward Recursion

We now define notation which allows us to solve formally the shortest-path problem. Let

$$d_{ij} = \text{the cost of transforming the system from state } i \text{ to state } j;$$
$$s_n = \text{the states of the } n\text{th stage;}$$
$$N = \text{the number of stages of the system} - 1;$$
$$f_n(s) = \text{the return of state } s \text{ when the system is at stage } n.$$

Using these definitions, the formal calculations for the solution of the problem are

$$f_0(s_1) = 0,$$

$$f_1(s_2) = f_0(s_1) + d_{12} = 0 + 16 = 16,$$

$$f_1(s_3) = f_0(s_1) + d_{13} = 0 + 14 = 14,$$

$$f_1(s_4) = f_0(s_1) + d_{14} = 0 + 12 = 12,$$

$$f_2(s_5) = \text{Min}\{f_1(s_2) + d_{25}, f_1(s_3) + d_{35}\}$$
$$= \text{Min}\{16+8, 14+14\} = 24,$$

$$f_2(s_6) = \text{Min}\{f_1(s_2) + d_{26}, f_1(s_3) + d_{36}, f_1(s_4) + d_{46}\}$$
$$= \text{Min}\{16+19, 14+20, 12+8\} = 20,$$

$$f_2(s_7) = \text{Min}\{f_1(s_3) + d_{37}, f_1(s_4) + d_{47}\}$$
$$= \text{Min}\{14+11, 12+17\} = 25,$$

$$f_3(s_8) = \text{Min}\{f_2(s_5) + d_{58}, f_2(s_6) + d_{68}\}$$
$$= \text{Min}\{24+20, 20+13\} = 33,$$

$$f_3(s_9) = \text{Min}\{f_2(s_5) + d_{59}, f_2(s_6) + d_{69}\} f_2(s_7) + d_{79}\}$$
$$= \text{Min}\{24+9, 25+17, 25+10\} = 33,$$

$$f_3(s_{10}) = \text{Min}\{f_2(s_6) + d_{6,10}, f_2(s_7) + d_{7,10}\}$$
$$= \text{Min}\{25+16, 25+20\} = 41,$$

$$f_4(s_{11}) = \text{Min}\{f_3(s_8) + d_{8,11}, f_3(s_9) + d_{9,11}, f_3(s_{10}) + d_{10,11}\}$$
$$= \text{Min}\{33+11, 33+12, 41+10\} = 44.$$

Backtracking

$$f_4(s_{11}) - d_{8,11} - f_3(s_8) = 44 - 11 - 33 = 0, \checkmark$$

$$f_4(s_{11}) - d_{9,11} - f_3(s_9) = 44 - 12 - 33 \neq 0,$$

$$f_4(s_{11}) - d_{10,11} - f_3(s_{10}) = 44 - 10 - 41 \neq 0,$$

$$f_3(s_8) - d_{58} - f_2(s_5) = 33 - 9 - 24 = 0, \checkmark$$

$$f_3(s_8) - d_{68} - f_2(s_6) = 33 - 17 - 25 \neq 0,$$

$$f_2(s_6) - d_{26} - f_1(s_2) = 20 - 19 - 16 \neq 0,$$

$$f_2(s_6) - d_{36} - f_1(s_3) = 20 - 20 - 14 \neq 0,$$

$$f_2(s_6) - d_{45} - f_1(s_4) = 20 - 8 - 12 = 0, \checkmark$$

$$f_1(s_4) - d_{14} - f_0(s_1) = 12 - 12 - 0 = 0. \checkmark$$

The shortest path is $1 \to 4 \to 6 \to 8 \to 11$ with length $= f_4(s_{11}) = 44$ km.

In our hiker's problem we performed the calculations in the most natural order, that is, by following the direction in which the hiker walks. Thus we calculated the return functions in the numerical order of their subscripts: f_0, f_1, f_2, f_3, and f_4. Such a "direction" of calculation is called *forward recursion*.

In general, the calculation for the return $f_n(s)$ at stage n of state s is

$$f_n(s) = \underset{s_{n-1}}{\text{Min}} \left\{ f_{n-1}(s_{n-1}) + d_{s_{n-1}s} \right\}, \qquad n = 1, 2, \ldots, N. \tag{2.38}$$

Equation (2.38) is the general forward-recursion equation and equations of this form are called *recursive equations*. They are used, in one form or another, to solve DP problems. The set of decisions which make up a problem solution is called a *policy*. A set of decisions which transforms the system from an intermediate stage to the final stage is called a *subpolicy*. We noticed earlier that if the hiker leaves state 5, the later trail choices are not affected by how the hiker got to state 5. Generalizing this statement and phrasing it in the above language, we can say that "an optimal subpolicy at any stage depends only upon that stage and the transformation costs from that stage, *not* upon choices made at earlier stages." This sums up the "principle of optimality of dynamic programming":

> The decisions of the optimal policy for stages beyond a given stage will constitute an optimal subpolicy regardless of how the system entered that stage.

It can be seen from (2.38) that $f_n(s)$ is a function of $f_{n-1}(s_{n-1})$ and $d_{s_{n-1}s}$ and this function is simply one of addition. In some problems, however, the function is more complicated. Consider, for example, the case where the transformation "costs" are probabilities and one is attempting to minimize the overall probability of a complete path from the initial to the final stage. Here the function between $f_{n-1}(s_{n-1})$ and $d_{s_{n-1}s}$ is one of multiplication. We state the general relationship by $\oplus$ (which may be addition, multiplication, or some other operation). The general formula for forward recursion is

$$f_n(s) = \text{Optimize}\{ f_{n-1}(s_{n-1}) \oplus d_{s_{n-1}s} \}, \qquad n = 1,2,3,\ldots, N. \quad (2.39)$$

We use the term "optimize" rather than "minimize" because in some problems we may wish to maximize the "costs" which are, in this case, benefits.

2.2.3. Backwards Recursion

As has already been mentioned, the values of the return functions for the shortest-path problem were calculated in the order of increasing subscripts. When the return functions are calculated in the order $f_N, f_{N-1}, \ldots, f_1$, the approach is termed *backwards recursion*. It usually means performing the calculations in the opposite direction to that of the events as they will occur when the solution is implemented. We now illustrate backwards recursion by using it to solve the hiker's problem. We must first redefine the return functions as

$f_n(s)$ = the return from the future stages $(n+1, n+2, \ldots, N)$ when the system is in state s at stage n.

In terms of our shortest-path problem, $f_n(s)$ is the shortest distance from hut s (state s) at stage n to the final stage. Thus $f_n(s)$ is the distance the hiker still has to walk. [This is usually different from the distance the hiker has already walked, which is what $f_n(s)$ was in forward recursion.]

Once hut 11 is reached there is no further distance to be walked so $f_4(s_{11}) = 0$. At the third stage only one further trail has to be walked, so the returns are simply their lengths. At the second stage we must calculate the minimum distance from each state to state 11 using the lengths of the trails from stage 2 and the stage 3 returns. Continuing in this way we are eventually able to establish that $f_0(s_1) = 44$. Naturally this is the same distance as calculated earlier. We can backtrack in a similar manner to that used for forward recursion to find the actual shortest path. Of course it is the same path as that found in Section 2.2.1. The complete set of calculations for backwards recursion are:

$$
\begin{aligned}
f_4(s_{11}) &= 0,\\
f_3(s_8) &= d_{8,11} + f_4(s_{11}) = 11 + 0 = 11,\\
f_3(s_9) &= d_{9,11} + f_4(s_{11}) = 12 + 0 = 12,\\
f_3(s_{10}) &= d_{10,11} + f_4(s_{11}) = 10 + 0 = 10,\\
f_2(s_5) &= \text{Min}\{d_{58} + f_3(s_8), d_{59} + f_3(s_9)\}\\
&= \text{Min}\{20 + 11, 9 + 12\} = 21,\\
f_2(s_6) &= \text{Min}\{d_{68} + f_3(s_8), d_{69} + f_3(s_9), d_{6,10} + f_3(s_{10})\}\\
&= \text{Min}\{13 + 11, 17 + 10, 16 + 10\} = 24,\\
f_2(s_7) &= \text{Min}\{d_{79} + f_3(s_9), d_{7,10} + f_3(s_{10})\}\\
&= \text{Min}\{10 + 12, 20 + 10\} = 22,\\
f_1(s_2) &= \text{Min}\{d_{25} + f_2(s_5), d_{26} + f_2(s_6)\}\\
&= \text{Min}\{8 + 21, 19 + 24\} = 29,\\
f_1(s_3) &= \text{Min}\{d_{35} + f_2(s_5), d_{36} + f_2(s_6), d_{37} + f_2(s_7)\}\\
&= \text{Min}\{14 + 21, 20 + 24, 11 + 22\} = 33,\\
f_1(s_4) &= \text{Min}\{d_{46} + f_2(s_6), d_{47} + f_2(s_7)\}\\
&= \text{Min}\{8 + 24, 17 + 22\} = 32,\\
f_0(s_1) &= \text{Min}\{d_{12} + f_1(s_2), d_{13} + f_1(s_3), d_{14} + f_1(s_4)\}\\
&= \text{Min}\{16 + 29, 14 + 33, 12 + 32\} = 44.
\end{aligned}
$$

Backtracking

$$
\begin{aligned}
f_0(s_1) - d_{12} - f_1(s_2) &= 44 - 16 - 29 \neq 0,\\
f_0(s_1) - d_{13} - f_1(s_3) &= 44 - 14 - 33 \neq 0,\\
f_0(s_1) - d_{14} - f_1(s_4) &= 44 - 12 - 32 = 0, \surd\\
f_1(s_4) - d_{46} - f_2(s_6) &= 32 - 8 - 24 = 0, \surd\\
f_1(s_4) - d_{47} - f_1(s_7) &= 32 - 17 - 22 \neq 0,\\
f_2(s_6) - d_{68} - f_2(s_8) &= 24 - 13 - 11 = 0, \surd\\
f_2(s_6) - d_{69} - f_3(s_9) &= 24 - 17 - 12 \neq 0,\\
f_2(s_6) - d_{6,10} - f_3(s_{10}) &= 24 - 16 - 10 \neq 0,\\
f_3(s_8) - d_{8,11} - f_3(s_{11}) &= 11 - 11 - 0 = 0. \surd
\end{aligned}
$$

The shortest path is $11 \leftarrow 8 \leftarrow 6 \leftarrow 4 \leftarrow 1$ with length $= s_1 = 44$ km.

In this example an equal amount of effort is required in order to solve the problem in either direction. However, there are problems for which the return function is more complicated than a simple arithmetical operation on two numbers. In some of these cases the two directions of computation require markedly differing amounts of computation, one from the other. The general backwards-recursive equation is

$$f_n(s) = \underset{s_{n+1}}{\text{Optimize}} \left\{ f_{n+1}(s_{n+1}) \oplus d_{s\, s_{n+1}} \right\}, \qquad n = N-1, N-2, \ldots, 0. \tag{2.40}$$

2.2.4. Exercises

1. Find the shortest path from the point with the lowest index to the point with the highest index for each of the networks in the exercises at the end of Chapter 3 using forward recursion.

2. Repeat problem 1 above using backwards recursion.

3. Repeat problems 1 and 2 above for the networks in Tables 3.1, 3.2, and 3.4.

4. Repeat problems 1, 2, and 3 above with $\oplus = \times$ (multiplication) rather than $+$(addition).

5. Solve the following problem by forward recursion:

$$\text{Maximize} \sum_{n=1}^{4} (6x_n - nx_n^2),$$

subject to

$$\sum_{n=1}^{4} x_n = 8,$$

$$x_n, \text{ nonnegative integer}, \qquad n = 1,2,3,4.$$

6. Solve problem 5 above by backwards recursion.

7. Compare the amount of computational effort required to solve problems 5 and 6 above.

8. A machine must produce 20 spindles over 4 days—Sunday (S), Monday (M), Tuesday (T), and Wednesday (W). Integer numbers of up to 6 can be produced on any day. It costs \$2 to store each spindle for any day after it is produced. The production costs for different numbers of spindles produced on the

different days are:

Number produced	Cost S	M	T	W
0	4	12	10	8
1	8	14	16	10
2	16	18	22	18
3	18	22	30	26
4	22	30	32	30
5	24	38	34	34
6	28	40	40	44

Solve this problem by forward recursion.

9. Solve problem 8 above by backwards recursion.

*10. A strong burglar can carry 100 kg in weight. The burglar finds seven objects to steal with values and weights:

$$(30, 50, 10, 20, 14, 25, 60) \quad \text{and} \quad (40, 50, 30, 10, 10, 40, 30),$$

respectively. Using forward recursion, decide which selection of objects is the most valuable to steal.

*11. Solve problem 10 above by backwards recursion.

*12. A farmer has 30 kg of pumpkin seed. If the farmer plants 1 kg of pumpkin seed, it will produce 3 kg of pumpkins in 1 year's time. The farmer has a very brave buyer who will guarantee to pay \$450, \$390, \$330, \$44, \$15, and \$5 for 1 kg of pumpkins over the next 5 years, respectively. The farmer decides to retire in 5-year's time and wants to dispose of all the pumpkins and seeds by the end of the fifth year. What is the farmer's optimal planting and selling policy if either must be done in integral amounts?
(a) Decide this by forward recursion.
(b) Decide this by backwards recursion.

2.3. Complexity

Combinatorial mathematics is often described as the study of the arrangement and selection of discrete objects. *Combinatorial optimization* (CO), is concerned with identifying the best possible arrangement or selection from among all those possible. As there is usually a finite number of possibilities for any given problem, it is theoretically possible to examine them all and choose the best. Unfortunately, there are just too many solutions to any nontrivial problem for this approach to be feasible. An example of this was given at the end of Chapter 0.

It is therefore of interest to attempt to design algorithms, which are more effective than complete enumeration. We turn now to the question of evaluating the effectiveness of an algorithm. The concept of effectiveness was placed on a firm scientific foundation by Edmonds (1965), whose work caused the following convention to be adopted by most of those concerned with algorithm efficiency:

> An algorithm is considered to be effective if it can guarantee to solve any instance of the problem for which it was designed by performing a number of elementary computational steps and the number can be expressed as a polynomial function of the size of the problem.

It is assumed that computational time is linearly proportional to the number of elementary computational steps required to implement the algorithm. The *size* of a specific instance of a problem is defined to be the number of symbols required to describe it.

It is a valid question to ask whether an effective, or "polynomial-time," algorithm can be devised for a given CO problem. There exist obscure CO problems in Number Theory for which it has been shown that no effective algorithm exists, and problems in other areas for which polynomial-time algorithms have been devised. This second class of problems is denoted by P (polynomial). Examples of problems in P are given in Sections 3.1 and 3.2.

There exists a third class of problems whose status is unknown. It is possible to devise algorithms for each problem, but no effective algorithm is known for any of them. However, neither has there yet appeared a proof showing that any are intractable. Our problem of finding the shortest path through a given set of cities lies in this last class which is denoted by NP (nondeterministic polynomial).

Within NP there is a subset of problems which is called NP-Complete. A problem is termed NP-Complete if it (1) belongs to NP and (2) has the property that if an effective algorithm is found for it then an effective algorithm can be found for every problem in NP. In this sense the NP-Complete problems are the hardest in NP. To establish the status of a CO problem for which no effective algorithm is known, it is usual to employ the concept of *reducibility*.

A problem p_1 is said to be reducible to problem p_2 (written $p_1 \propto p_2$) if the existence of an effective algorithm for p_2 implies the existence of an effective algorithm for p_1. The following result is often used to establish that a problem $p \in$ NP is NP-Complete.

Theorem 2.1. *If p_1 is NP-Complete and $p_1 \propto p_2$ then p_2 is also NP-Complete.*

(For a proof, see Garey and Johnson, 1979, p. 38.)

Many of the problems in NP have defied the attempts to find effective algorithms of some of the best mathematicians over the past 30 years. There is also more objective circumstantial evidence that P ≠ NP. Thus it seems unlikely that an effective algorithm exists for any of the NP-Complete problems. Hence the fact that a problem is NP-Complete is considered justification for heuristic procedures to be applied to it, that is, procedures which do not guarantee to produce an optimal solution for every instance of the problem. The challenge is to find heuristics with good-performance guarantees which are also effective in the sense defined earlier. Heuristics are discussed in the next section.

2.4. Heuristic Problem Solving

An *algorithm* for a problem is a scientific procedure which will converge to the best feasible solution to the problem. Analysts in business and industry are often faced with problems of such complexity that the standard algorithms are inappropriate. There are many reasons why this might be so.

(1) The dimensions of the problem may be so large that the application of the fastest-known algorithm on the fastest computer may take a prohibitive amount of computational time. This is certainly true for certain vehicle routing problems.
(2) The problem may be virtually impossible to formulate in explicit terms. The aims of different managers involved in operating a system may be conflicting or ill-defined. In fact it may be difficult to express many features of the problem in quantitative terms.
(3) Data collection may be beset with problems of accuracy and magnitude. For example, in large-scale location problems the analyst may be faced with calculating an enormous number of location-to-location distances. In order to provide this information in reasonable time it may be necessary to make approximations. Sometimes the use of approximate data makes the concept of an optimal solution meaningless.

Of course, a manager will find no comfort at all in a consultant saying that the literature does not contain an efficient method which guarantees an optimal solution for the problem in reasonable computational time. Come what may, the manager somehow has to schedule flight crews, route delivery vans, or whatever. At this point the analyst has a number of options:

(1) Develop the methodology that will provide optimal solutions efficiently.
(2) Find algorithms that will solve certain special cases of the problem.
(3) Look for efficient algorithms that solve a relaxed version of the problem.
(4) Come up with algorithms that seem likely to run quickly most, but not all, of the time.
(5) Give up the quest for optimality and provide approximate methods that run quickly but have no guarantee of optimality.

Aim (1) is often unrealistic. Aims (2) and (3) are occasionally appropriate. However, there is a very real danger: People have warned of the pitfalls of substituting the real problem with an artificial one which corresponds to standard models and known techniques. Nothing can give consultancy a worse name than the consultant who "bends" a clients' problem into a form which is amenable to solution by the consultant's pet method, producing solutions which are of little practical use. However aim (4) is obviously worthy. It is usually satisfactory to employ an algorithm which almost always runs in reasonable time. The simplex algorithm for linear programming (see Chapter 1) is a good example. However, if such a method cannot be found we are left with aim (5).

The idea of approximate methods, which are easy to use but which give up the guarantee of optimality, is not new. Indeed as early as 300 A.D. Pappas, writing on Euclid, suggested this approach. Descartes and Leibnitz both attempted to formalize the subject. It became known as the study of *heuristics* and *heuristic* was the name of an area of academic study whose aim was to investigate the methods of discovery and invention. It was allied with logic, philosophy, and psychology. The name itself was derived from the Greek word *heuriskein*—to discover. Today the term *heuristic* is used to describe a method "which, on the basis of experience or judgment, seems likely to yield a good solution to a problem, but which cannot be guaranteed to produce an optimum."

Most people use heuristics all the time in their day-to-day lives:

- In order to guard against catching a cold, if clouds are building up in the West, take a raincoat.
- In order to guarantee never to run out of gas, when the gasoline gauge shows less than one-quarter full, buy more gas for the car.
- In order to avoid problems with one's bank manager, transfer funds from the savings account to the checking account when the latter is in the red by more than $100.

Heuristics abound in business too:

- Reorder when only one-third of the stock is left.
- Schedule the urgent jobs first.
- Allocate 4% of last year's sales revenue to advertising.

There are four basic strategies for heuristic procedures. Many methods comprise a combination of more than one of these strategies.

1. *The Construction Strategy*. The input for methods based on this strategy is nothing more than the data which define a specific instance of the problem. A solution is built up one component at a time. A construction strategy begins by examining these data and attempting to identify an element of the final solution which is likely to be a valuable part of a very good final solution. Next, successive additional elements of a solution are

added. The better construction heuristics employ some kind of "look-ahead" mechanism. That is, additions are made not just because they appear as a good idea at the time, but because they are likely to be of genuine value in the complete solution. Once the final solution has been built up it may be obvious that improvements can be easily effected. Thus the strategy is often applied to the output of the construction method. This strategy is worthwhile when it is relatively difficult to generate feasible solutions to the problem.

2. *The Improvement Strategy*. The input for methods based on this strategy is a solution to the problem. This solution is then progressively improved by a series of modifications. In some instances it may be impossible to make much progress in this way and yet the final product may still be far from optimal. Some improvement strategies are farsighted in the following sense. Some iterations of the "improvement" process may actually be allowed to bring about a temporary worsening in solution value if it can be seen that this will create a situation where worthwhile gains can eventually be made. This strategy is useful when it is relatively easy to generate starting solutions. A variety of solutions can be used as input and the best final result chosen. Sometimes the strategy is used to convert an infeasible solution into a feasible one.

3. *The Component Analysis Strategy*. Some problems are so large or so complicated that the only practical approach is to break them up into manageable portions. Sometimes these portions are then dealt with independently by heuristics or algorithms. The solutions for the portions are then joined to form some master plan. Of course, it may be extremely difficult to piece together the solutions to the different components into an acceptable plan. If the components can be ordered in some sort of logical sequence it usually makes sense to examine them in the same order. This ordering is often based on some time scale which is an integral part of the problem. For instance, in investment problems a natural sequencing may be to define a component as 1 year's activities. Obviously, the output of the analysis of one component may be a valuable input for the analysis of later components.

4. *The Learning Strategy*. Methods based on this strategy often use a tree-search diagram to chart their progress. That is, the different options which appear at various stages are represented by different branches of a tree. The sequences of choices actually made can be traced by a path through the tree. The choice of which branch to take is guided by learning from the outcome of earlier decisions. The early termination of a branch-and-bound search is an example of this strategy.

We turn now to the problem of how to design an effective heuristic.

2.4.1. The Design of Heuristics

When confronted with the problem of designing a heuristic for a specific problem it is often very useful to set aspiration levels before plunging into the actual design process; that is, what type of performance guarantees must the heuristic possess? The guarantees must apply to the actual problems under study. This usually means large-scale problems, not just small test problems. Desirable characteristics include:

(1) Execution in reasonable computational time.
(2) Solutions which are close to optimality on the average.
(3) Only a small probability of any one solution being far below optimality.
(4) Simplicity of both design and computational requirements.

Characteristics (2) and (3) suggest an average-case performance bound be calculated.

Returning to the point of guaranteeing good performance on realistic and not just small test problems we turn to the question of validation. Sometimes this can be achieved by constructing large-scale problems with known optimal solutions. That is, this solution is defined and then the rest of the data for the problem are "built around it" ensuring the optimality of S. Although such problems are inherently artificial, careful construction can provide useful test problems. Despite the previous remark, it is often still quite useful to compare heuristic performance with that of an algorithm on a wide variety of small- and medium-sized test problems if possible.

When validation is difficult but it is relatively easy to generate feasible solutions, the following design approach may be valuable. Called the better-than-most approach (BTM), it randomly generates solutions, selecting those that are feasible. The best solutions among these are selected according to the optimality criteria of the analyst, where some of these criteria may be difficult to quantify. There is the obvious question of how large a sample must be taken in order to guarantee that the best solution identified will be with a specified probability, within a given percentage of optimality.

This approach leads to the question of when to terminate a heuristic that generates one or more feasible solutions early in its execution and then spends most of its time searching for improvements.

When optimality criteria are difficult to quantify, it may be reasonable to contemplate the design of a heuristic which involves interaction with an analyst during its execution. Using intuition and experience, the analyst may be able to guide the heuristic search process far more efficiently than a predetermined set of rules. There is the further strategy of relaxing certain difficult constraints and having the analyst choose from a variety of solutions proposed by the interactive program. The analyst then, on the basis of his or her choice, may be able to set slightly more constrained problems for the interactive heuristic. These approaches have been successfully used to solve large-scale vehicle scheduling problems heuristically.

2.4.2. The Use of Heuristics

Sometimes good heuristics have obvious advantages over the more standard algorithms of Combinatorial Optimization:

(1) They can often be implemented by a practitioner who lacks advanced training, by virtue of their simplicity.
(2) They require modest computational time, thus representing a significant saving in cost.
(3) They are flexible, often allowing use as planning tools whcre many scenarios and sets of nonquantifiable constraints must be considered.

The obvious disadvantage is the fear that solutions far below optimality will be produced. It is often not clear how significant the deficiency is. There can be further problems with implementation. Most of the literature on the topic deals with successfully conceptualizing heuristic solutions for idealized models. (It is a shame that more negative results are not reported as these often contribute to the learning process of others as well.) Many of the heuristics posed do not pass the test of being subjected to real-world problems because of model deficiency. There are often complications not present in the model for which the heuristic is designed. Thus successful heuristic implementation often lags behind heuristic development by a considerable degree.

In order to safeguard against the error of introducing a heuristic at the expense of existing or new standard techniques, the management should ask the following questions:

(1) Will we spend more than we save by developing and implementing a new heuristic?
(2) Will the decision process be unnecessarily slowed down by its introduction?
(3) Is there a chance that work quality will be reduced to a level that negates the savings brought about by its introduction?
(4) Are the results, when actually implemented, not really significantly better than those produced before its introduction?

If the answer to any of these questions is in the affirmative the manager should seriously question the value of the introduction of the heuristic proposed.

CHAPTER 3

Optimization on Graphs and Networks

3.1. Minimal Spanning Trees

Natural gas is soon to be available in Scenic Valley (SV). There are 10 towns in the SV area. One town, Geyser Town (GT), contains the wellhead. The SV Gas Company (SVGC) has the task of designing a supply system so that there is a path of pipes from GT to every other town. The manager of SVGC wishes to lay the smallest amount of pipe possible. His team of engineers have drawn up Table 3.1 showing the length of pipe (in kilometers) required for a connection between each pair of towns. Connections at points other than towns are prohibited. This is shown in Fig. 3.1, based on a map of SV.

Table 3.1

		2	3	4	5	6	7	8	9	10
(Geyser Town)	1	8	5	9	12	14	12	16	17	22
(Lakeside)	2		9	15	17	8	11	18	14	22
(Mountainview)	3			7	9	11	7	12	12	17
(Riverview)	4				3	17	10	7	15	18
(Twin Streams)	5					8	10	6	15	15
(Snowdonia)	6						9	14	8	16
(Skier's Paradise)	7							8	6	11
(Park City)	8								11	11
(Rolling Downs)	9									10
(Tumbling Basin)	10									

The minimum length layout can easily be constructed using one of the well-known methods of Kruskal (1956) and Prim (1957). We begin with Kruskal's method shortly, but first we examine the problem a little more deeply.

It is clear that GT need not be distinguished just because it contains the wellhead. This is simply because we wish to create a network of pipes connecting all the towns. Which town contains the wellhead is irrelevant in this endeavor.

We can model the problem in terms of graph theory. (See Chapter 5, Section 5.2 for an explanation of the terminology used here.) The problem has a graph theoretic equivalent in which each town is represented by a vertex and each potential pipeline by an edge. The problem becomes one of finding a spanning tree of the resulting graph with minimum total edge weight—a minimal spanning tree. Note that a tree does not contain a cycle so that the length of any solution with a cycle of pipes can be reduced by removing any pipe in the cycle. We now return to our original problem and solve it by Kruskal's method.

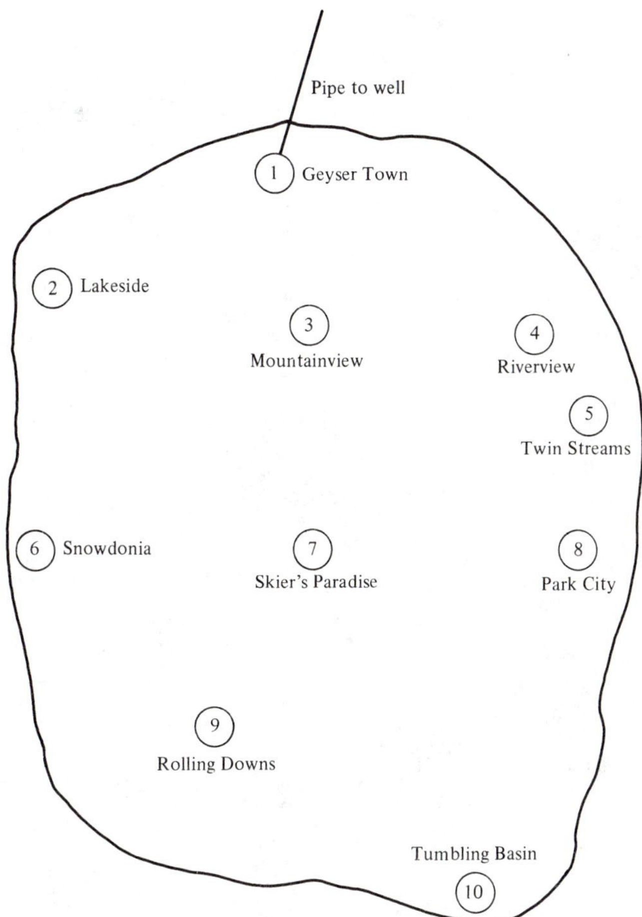

Figure 3.1. Map of 10 Towns in Scenic Valley.

3.1.1. Kruskal's Method

We begin by scanning Table 3.1 and finding the pipe of least length. This joins Riverview (4) to Twin Streams (5) and has length 3. This pipe is then accepted as a part of the solution. (The reader may find it beneficial to reproduce Fig. 3.1 and add the pipelines to that diagram as the solution is built up.) The next shortest pipe length is then identified—it joins town 1 to town 3 with length 5. It is accepted as part of the solution. In selecting the next largest pipeline we have a choice: 5–8 and 7–9, both of length 6. In this case an arbitrary choice is made, say 5–8, and this pipe is accepted as part of the solution. The next longest pipe length is naturally 7–9 and it is also accepted. At the next stage there are a number of choices: 3–4, 3–7, and 4–8, all of length 7. If the first of these is chosen arbitrarily it can be accepted, as well as the second. However, the third choice, 4–8, creates a cycle of pipes: $\langle 4,5,8,4 \rangle$. We know that the minimum solution will not contain a cycle, so we reject 4–8. At the next stage there are even more choices: 1–2, 2–6, 6–9, 5–6, and 7–8, all of length 8. If all the first three are accepted, the cycle $\langle 1,2,6,9,7,3,1 \rangle$ will be created. Suppose that 1–2 and

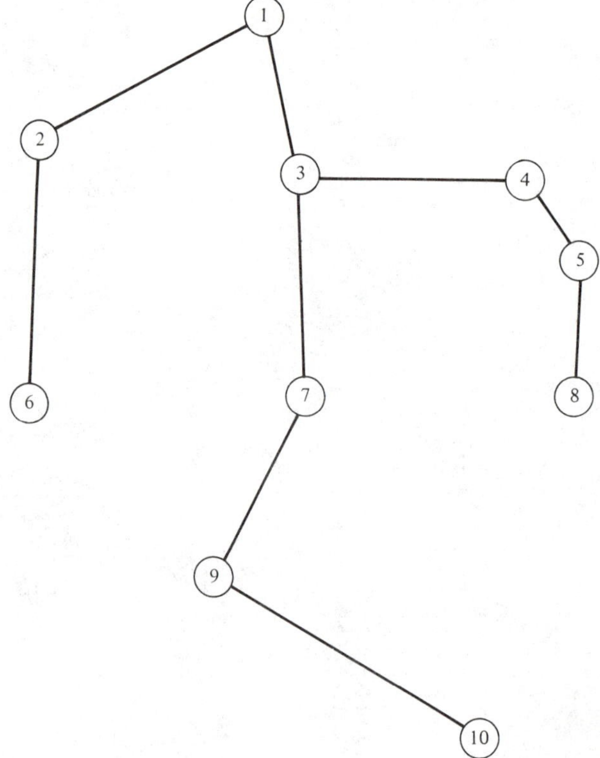

Figure 3.2. A Minimal Network of Pipelines Derived by Kruskal's Method.

2–6 are chosen arbitrarily, and 6–9 is rejected; 5–6 and 7–8 are also rejected as they create cycles. Moving on to the next stage we see that all the choices of length 9, 1–4, 2–3, 3–5, and 6–7, create cycles. Hence they are all rejected. Finally, we accept pipe 9–10 of length 10 and the network is complete. The network is shown in Fig. 3.2 and has total length

$$3+5+6+6+7+7+8+8+10=60 \text{ km}.$$

The procedure illustrated above is now generalized.

Kruskal's Method

Objective: To find the minimum weight spanning tree T, of a weighted graph G with n vertices.

Steps

(1) Assemble the edges of G in order of nonincreasing edge weight.
(2) Examine each edge of G in the order created in (1). Accept this edge as part of the solution unless it creates a cycle with the edges already accepted. Settle ties arbitrarily.
(3) Terminate when $n-1$ edges have been accepted.

3.1.2. Prim's Method

We explain Prim's method by using it to solve the gas reticulation problem just discussed. As in Kruskal's method we begin by accepting the pipe of least length, which is 4–5 of length 3. The two towns incident with this pipe, 4 and 5, belong to what is called a *component* (in graph theoretic terms, a *connected subgraph*) (see Section 5.2 for further information on Graph Theory). We now find the town not in the component which is closest (requiring the least-cost pipe to connect) to a town in the component. That is, we consider towns 4 and 5 as one unit and find the town closest to them. Table 3.1 reveals that it is town 8, which is 6 km from town 5. Pipe 8–5 is accepted as part of the solution. (Once again the reader may care to reproduce Fig. 3.1 and draw in the accepted pipes as we proceed.) This creates a component with towns 4, 5, and 8. We repeat this step and look for the closest town to one of these three. It is town 3 which is 7 km from town 4. Pipe 3–4 is accepted. Note that, unlike Kruskal's method, the pipe 4–8 is not considered because it links two towns already in the component. Town 3 is added to the component which now contains 3, 4, 5, and 8.

Continuing this process we see that the next town to be included in the component is town 1 which is 5 km from town 3. Pipe 1–3 is accepted. Next

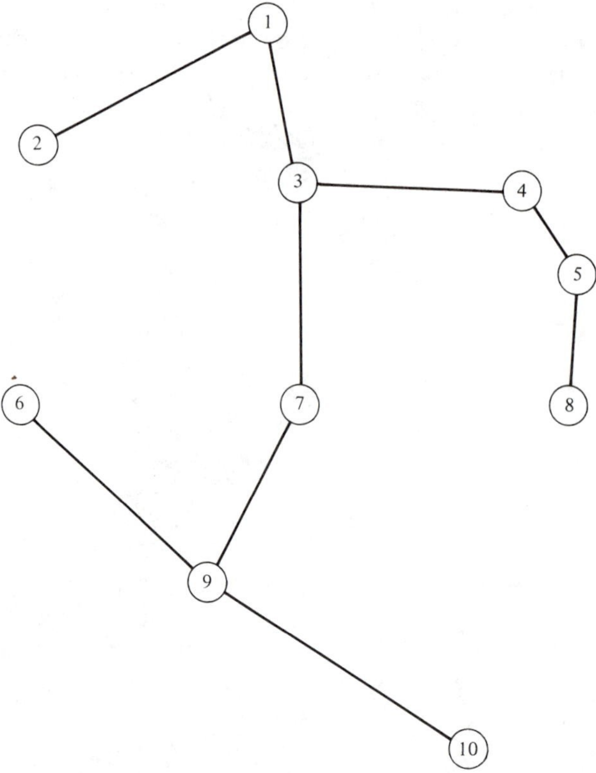

Figure 3.3. A Minimal Network of Pipelines Derived by Prim's Method.

town 7 is included as it is 7 km from town 3. Pipe 3–7 is accepted. Then town 9, which is 6 km from town 7, is included. Pipe 7–9 is accepted and the component contains towns 4, 5, 8, 3, 1, 7, and 9. At the next iteration we see that towns 2 and 6 are both 8 km from towns 1 and 9, respectively. This is the first tie and it is settled arbitrarily. Pipe 1–2 is accepted and town 2 is included in the component. Pipe 6–9 can also be accepted and town 6 is included. This leaves only town 10 not in the component. It is closest to town 9 so that pipe 9–10 is accepted and the component contains all the towns. When this occurs the network is complete. It is shown in Fig. 3.3 and has a different structure from that derived by Kruskal's method in Fig. 3.2. It has total length

$$3+6+7+5+7+6+8+8+10=60 \text{ km}.$$

The procedure illustrated above is now generalized.

Prim's Method

Objective: To find the minimum weight spanning tree T of a weighted graph G with n vertices in set P and edge weight d_{ij} for each edge ij.

Steps

(1) Define the set of vertices in a component of vertices to be built up to be C. Set

$C = \{i, j\}$, where edge ij is the edge of least weight;

$A = \{ij\}$, where A is the set of edges which are currently part of the final spanning tree.

(2) For all $i \in P \backslash C$, find a vertex $k_i \in C$ such that

$$d_{ik_i} = \operatorname*{Min}_{j \in C} \{d_{ij}\} = \alpha_i.$$

Set i to have a label $[k_i, \alpha_i]$. If no such label can be found, set i to have label $[0, \infty]$.

(3) Identify i^* such that

$$\alpha_{i^*} = \operatorname*{Min}_{k_i \in P \backslash C} \{\alpha_{k_i}\}.$$

Set

$$C \leftarrow C \cup \{i^*\}, \qquad A \leftarrow A \cup \{k_{i^*} i^*\}.$$

If $|C| = n$, terminate. Otherwise, continue.

(4) For all $i \in P \backslash C$ such that i^*i is an edge in G and $\alpha_i > d_{i^*i}$, set $\alpha_i = d_{i^*i}$ and $k_i = i^*$. Go back to step (3).

A spanning tree on n vertices always contains $n-1$ edges. Thus a solution to our problem always contains $n-1$ pipes. The solutions generated by the two methods as shown in Figs. 3.2 and 3.3 are different. However, either solution could have been attained by the other method by settling ties in a different manner. Indeed both methods guarantee to find minimum solutions to every problem to which they are applied—they are both algorithms not heuristics. However, Prim's method sidesteps the problems of having to detect cycles, and having to order all the edges in terms of their weight. Thus it is, in general, more efficient than Kruskal's method.

3.1.3. Exercises

1. Find the minimal spanning tree for the following table of distances using Kruskal's method:

	1	2	3	4	5	6	7	8	9	10
2	3									
3	4	16								
4	9	19	13							
5	8	20	1	16						
6	7	4	2	4	13					
7	6	12	5	9	3	6				
8	5	14	9	7	14	7	4			
9	4	17	14	6	5	9	5	7		
10	20	8	15	11	9	3	19	9	10	
11	15	3	20	12	19	12	6	19	15	20

2. If d_{12} is increased from 3 to 20 in problem 1, how is the optimal tree affected?
3. Solve problem 1 above using Prim's method.
4. Solve problem 2 above using Prim's method.

*5. Devise an efficient method for solving problems of the form of 2 above.

*6. Prove that any tree spanning n vertices contains $n-1$ edges.

7. Prove that any uniquely shortest edge of a weighted graph G must be part of any minimal spanning tree of G.

*8. Suppose that any chord c (an edge not part of a spanning tree) is added to a minimal spanning tree T of a graph G. Prove that c has length no less than that of any edge in the cycle its addition creates.

3.2. Shortest Paths

The tourist area of Scenic Valley, introduced in Section 3.1, has a network of aircraft flights. The times (in minutes) denoted by c_{ij} to travel between towns i and j are given in Table 3.2. The times are slow because they include time for boarding and disembarking. A "∞" indicates that it is not possible to travel directly between the various towns. A diagram of the possible flight segments is given in Fig. 3.4.

Geyser Town (GT) is the only town whose airport offers flights to cities outside the valley. Thus all tourists flying in arrive at GT and then proceed to the other valley centers. The SV Tourist Agency has to solve repeatedly the question: Which is the quickest route from GT to the various other towns? The tourist agent decides to solve the problem once and for all by calculating the nine required shortest paths—from GT to each of the other towns. This can be accomplished efficiently by Dijkstra's method (1959) which is done in the next section. (A method for finding just one of the shortest paths, using Dynamic Programming, was explained in Chapter 2.)

Table 3.2

		2	3	4	5	6	7	8	9	10
(GT)	1	80	50	90	∞	∞	∞	∞	∞	∞
	2		90	∞	∞	80	110	∞	∞	∞
	3			70	∞	∞	70	120	∞	∞
	4				30	∞	∞	70	∞	∞
	5					∞	∞	60	∞	∞
	6						90	∞	80	∞
	7							80	60	110
	8								∞	110
	9									100

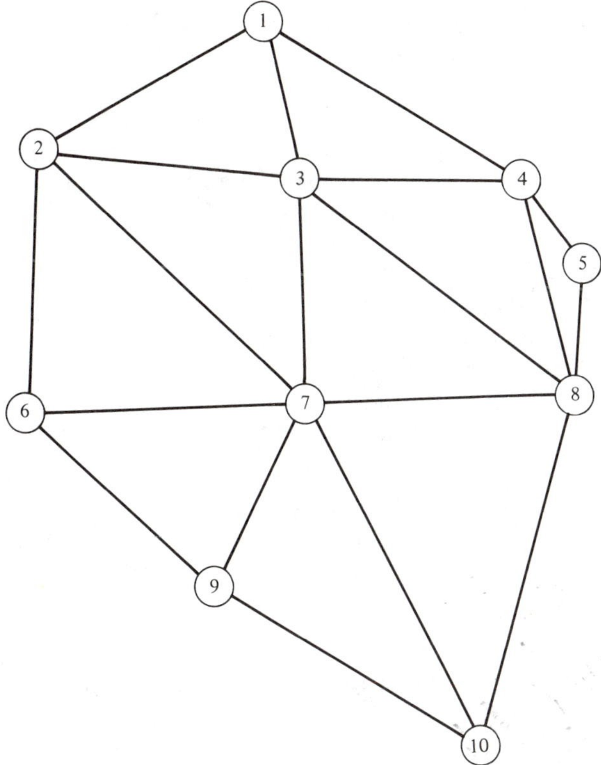

Figure 3.4. The Available Flight Segments in Scenic Valley.

3.2.1. Dijkstra's Method

In order to implement the method, a label $D(i)$ is defined for each town i as the shortest time required to fly from GT to i. Naturally we set $D(1)=0$ as GT is town 1. The labels for the other towns are as yet undefined. We begin by finding the town which is closest to GT. A scan of the first row in Table 3.2 reveals that it is town 3, which is 50 min away. Thus we can set $D(3)=50$. The method then calculates the other labels, one for each iteration. When all the towns are labeled, it calculates the actual shortest paths which go with these shortest times. The general step of the labeling process is as follows. Identify each flight segment which joins a labeled town to an unlabeled town. At this instant, as towns 1 and 3 are labeled, such flight segments are 1–2, 1–4, 3–2, 3–4, 3–7, and 3–8. The reader will find it beneficial to make a copy of Fig. 3.4 and label the towns as indicated. We now calculate the various possibilities for shortest times involved with these segments. The 1–2 segment implies that we could fly from town 1 to town 2 in $D(1)+c_{12}=0+80=80$ min. Similarly, $D(1)+c_{14}=0+90=90$ min.

From the first calculation, we know that we can fly to town 2 in no more than 80 min. Further, $D(3)+c_{32}=50+90=140$, $D(3)+c_{34}=50+70=120$, $D(3)+c_{37}=50+70=120$, and $D(3)+c_{38}=50+120=170$.

The smallest time produced by all these calculations is the first and $D(2)$ is set to 80. The general step is repeated and the reader should check that it leads to setting $D(4)=90$. In successive iterations the following labelings occur: $D(5)=120$, $D(7)=120$, $D(6)=160$, $D(8)=180$, $D(9)=180$, and $D(10)=230$.

Having labeled all the towns, a backtracking procedure is invoked to find the actual shortest paths themselves. We know that the minimum time required to fly from town 1 to town 10 is 230 min. This value of $D(10)=230$ was arrived at by adding to the label of a neighboring town the time of the appropriate leg. That is, the path from 1 to 10 of 230 min travels through exactly one of 7, 8, and 9. Which of the following is true:

$$D(9)+c_{9,10}=D(10)?$$

$$D(8)+c_{8,10}=D(10)?$$

$$D(7)+c_{7,10}=D(10)?$$

As the last equation is the only one which is true, the shortest path to 10 must pass through 7. Repeating this process for town 7, we find that $D(3)+c_{37}=D(7)$. Hence the shortest path passes through town 3 as well. The complete 1–10 path is 1–3–7–10 with a time of $D(10)=230$. The complete list is given in Table 3.3.

Table 3.3

Town	Path	Time (min)
1	1	$D(1)=0$
2	1–2	$D(2)=80$
3	1–3	$D(3)=50$
4	1–4	$D(4)=90$
5	1–4–5	$D(5)=120$
6	1–2–6	$D(6)=160$
7	1–3–7	$D(7)=120$
8	1–3–8	$D(8)=170$
9	1–3–7–9	$D(9)=180$
10	1–3–7–10	$D(10)=230$

The procedure illustrated above is now generalized.

Dijkstra's Method

Objective: To find the shortest path from one given vertex in a weighted graph G, with vertex set P and edge weight c_{ij}, to all other vertices of G.

Steps

Denote the given vertex by p:

(1) Define $D(i)$ as the shortest distance from p to i yet known, for all $i \in P$. Set $D(p)=0$. Mark this label of p as permanent. Set $D(i)=\infty$ for all $i \in P \setminus \{p\}$. Mark these labels as temporary. Let A equal the set of vertices i in P for which $D(i)$ has been defined. That is, A is the set of vertices labeled so far with permanent labels. Let $B = P \setminus A$. Set $A = \{p\}$.

(2) If $B = \varnothing$ proceed to step (4). Otherwise continue. Update the label of each vertex j in B as

$$D(j) = \underset{i \in A}{\mathrm{Min}} \{ D(j), c_{ij} + D(i) \}.$$

Let k be such that

$$D(k) = \underset{j \in B}{\mathrm{Min}} \{ D(j) \}.$$

(3) $A \leftarrow A \cup \{k\}$ and $B \leftarrow B \setminus \{k\}$. Mark k with $D(k)$ as its permanent label. Go back to step (2).

(4) List all pairs i and $j \in P$ for which

$$D(j) = c_{ij} + D(i). \qquad (*)$$

Each edge ij for which (*) holds is on the shortest path from p to j.

(5) Terminate.

3.2.2. Remarks

The Dijkstra procedure can be terminated early if some, but not all, shortest paths from p are required. Once the labels for all the points in the given subset have been calculated, step (4) can be performed to find the actual paths.

The identification of the actual paths relies on the property that if the shortest path from p to j passes through i, then the $p-i$ part of the $p-j$ path constitutes a shortest path from p to i.

Steps (2) and (4) may create a number of ties. That is, there may be a number of different edges which all correspond to the minimum value sought. The method breaks ties arbitrarily, i.e., makes a choice at random. The existence of ties implies that there are alternative shortest paths.

3.2.3. Floyd's Method

There is another problem that the SV Tourist Agency has to solve. Having flown into a town via GT, tourists often want to drive between the first four towns in the valley in the shortest time. The driving times (in 10-min units) are given in Table 3.4. The roads that are available are given in Fig. 3.5.

Table 3.4

	1	2	3	4
(GT) 1	0	48	30	54
2	42	0	40	∞
3	36	60	0	91
4	48	∞	48	0

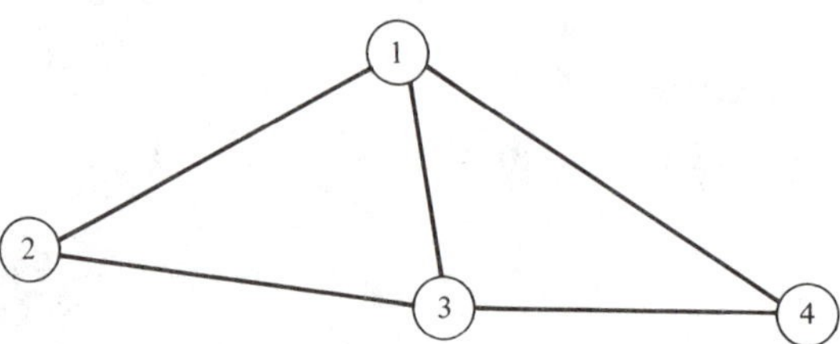

Figure 3.5. Roads between the First Four Towns in Scenic Valley.

The times are not always proportional to distance (as are the entries in Table 3.2) because some roads are freeways and others are not. Indeed it is sometimes quicker to travel via intermediate towns than by the direct route. Further, the time to go from a to b is often different from the time for the return journey because of the topography of the valley. The agency requires a set of shortest time paths between each pair of towns. Dijkstra's method, explained in the last section, could be used repeatedly with each town successively replacing GT. This is an inefficient way to tackle the problem as we are using the method in a way for which it was not designed. This leads to each shortest path being calculated twice—once in each direction. Floyd's method (1962), which is now explained, is a better way to go about the task.

In order to implement the method, five 4×4 matrices $\mathbf{D}^k$, $k=0,1,2,3,4$, are set up. Note that k is not a power but a superscript. The $i-j$ entry d_{ij}^k of $\mathbf{D}^k$ is defined to be the length of the shortest path from town i to town j with intermediate vertices on the path chosen from the first k towns $1,2,\ldots,k$. Naturally $\mathbf{D}^0$ is just Table 3.4, as these are the direct distances with no intermediate towns on the paths. One of our aims is to calculate $\mathbf{D}^4$, whose entries are the lengths of the shortest paths with any of the towns as intermediates. The other aim is to find the paths themselves, as before. For this we define another 4×4 matrix $\mathbf{P}$. The $i-j$ entry p_{ij} of $\mathbf{P}$ is defined to be the second town on the shortest path from town i to town j. The entries in $\mathbf{P}$ are updated as the method proceeds and they are not all correct according to the definition of p_{ij} until the method is terminated.

The $\mathbf{D}$ matrices are calculated in the order: $\mathbf{D}^0,\mathbf{D}^1,\mathbf{D}^2,\mathbf{D}^3,\mathbf{D}^4$. We shall not know all the correct entries of $\mathbf{P}$ until $\mathbf{D}^4$ is calculated. We begin by defining **P** as

$$\mathbf{P} = \begin{pmatrix} 1 & 2 & 3 & 4 \\ 1 & 2 & 3 & 4 \\ 1 & 2 & 3 & 4 \\ 1 & 2 & 3 & 4 \end{pmatrix}.$$

The rationale for this is that we do not know initially any of the shortest paths. So to start the process we assume that each shortest path from town i to town j is simply the direct road joining i to j. Thus the $i-j$ entry in **P** is j for each $i-j$ pair because it is defined to be the second town on the path.

The method begins by calculating $\mathbf{D}^1$ from $\mathbf{D}^0$ and updating **P** in the process. Recall that the $i-j$ entry in $\mathbf{D}^1$ contains the shortest time to travel from town i to town j, possibly via town 1. Naturally the first row (and the first column) of $\mathbf{D}^1$ will be identical to the first row (and the first column) of $\mathbf{D}^0$. The first entry that can possibly differ is d_{23}^1. Is it worthwhile to go through town 1 when traveling from town 2 to town 3? We have to compare $d_{23} = 40$ with $d_{21} + d_{13} = 42 + 30 = 72$. (The superscript 0 is dropped and d_{ij}^0 is written as d_{ij}.) We find that it is not worthwhile and $d_{23}^1 = d_{23} = 40$. We now calculate d_{24}^1. From Table 3.4, $d_{24} = \infty$, indicating that there is no direct road from town 2 to town 4. However, $d_{21} + d_{14} = 42 + 54 = 96$. Hence we set $d_{24}^1 = 96$. We must now update **P**. The second town on the shortest known path from town 2 to town 4 is no longer town 4 but town 1 as the shortest known path is 2–1–4. Thus p_{24} is updated to become 1.

Calculating d_{32}^1, we find that $d_{32} = 60$ and $d_{31} + d_{12} = 36 + 48 = 84$, so $d_{32}^1 = 60$ and no change is made to **P**. Calculating d_{34}^1, we find that $d_{34} = 91$ and $d_{31} + d_{14} = 36 + 54 = 90$. As we can shave a mile off the journey from town 3 to town 4 by passing through town 1, we set $d_{34}^1 = 90$ and set $p_{34} = 1$. The reader should verify that in calculating the fourth row of $\mathbf{D}^1$, $d_{42}^1 = 96$, $d_{43}^1 = 48$, and p_{42} is updated to 1. Thus

$$\mathbf{D}^1 = \begin{pmatrix} 0 & 48 & 30 & 54 \\ 42 & 0 & 40 & 96 \\ 36 & 60 & 0 & 90 \\ 48 & 96 & 48 & 0 \end{pmatrix}$$

and

$$\mathbf{P} = \begin{pmatrix} 1 & 2 & 3 & 4 \\ 1 & 2 & 3 & 1 \\ 1 & 2 & 3 & 1 \\ 1 & 1 & 3 & 4 \end{pmatrix}.$$

We now calculate $\mathbf{D}^2$ from $\mathbf{D}^1$. The $i-j$ entry of $\mathbf{D}^2$ contains the shortest time to travel from town i to town j by possibly passing through *both* towns 1 and 2. The first entry that can possibly be updated is d_{13}^2. Now $d_{13}^1 = 30$. But $d_{12}^1 + d_{23}^1 = 48 + 40 > 30$. Hence $d_{13}^2 = 30$. Continuing in this way, set

$$d^1_{14} = 54, \qquad d^1_{12} + d^1_{24} = 48 + 96, \quad \text{and} \quad d^2_{14} = 54.$$

The second row (and second column) of $\mathbf{D}^2$ must equal the appropriate entries in $\mathbf{D}^1$ and so do not need to be checked:

$$\begin{aligned}
d^1_{31} &= 36, & d^1_{32} + d^1_{21} &= 60 + 48, & \text{set } d^2_{31} &= 36;\\
d^1_{34} &= 90, & d^1_{32} + d^1_{24} &= 60 + 96, & d^2_{34} &= 90;\\
d^1_{41} &= 48, & d^1_{42} + d^1_{21} &= 96 + 42, & d^2_{41} &= 48;\\
d^1_{43} &= 48, & d^1_{41} + d^1_{13} &= 48 + 30, & d^2_{43} &= 48.
\end{aligned}$$

Thus

$$\mathbf{D}^2 = \begin{pmatrix} 0 & 48 & 30 & 54 \\ 42 & 0 & 40 & 96 \\ 36 & 60 & 0 & 90 \\ 48 & 96 & 48 & 0 \end{pmatrix}$$

and

$$\mathbf{P} = \begin{pmatrix} 1 & 2 & 3 & 4 \\ 1 & 2 & 3 & 1 \\ 1 & 2 & 3 & 1 \\ 1 & 1 & 3 & 4 \end{pmatrix}.$$

We now calculate $\mathbf{D}^3$ from $\mathbf{D}^2$:

$$\begin{aligned}
d^2_{12} &= 48, & d^2_{13} + d^2_{32} &= 30 + 60, & \text{set } d^3_{12} &= 48;\\
d^2_{14} &= 54, & d^2_{13} + d^2_{34} &= 30 + 40, & d^3_{14} &= 54;\\
d^2_{21} &= 42, & d^2_{23} + d^2_{31} &= 40 + 36, & d^3_{21} &= 42;\\
d^2_{24} &= 96, & d^2_{23} + d^2_{34} &= 40 + 90, & d^3_{24} &= 96,\\
d^2_{41} &= 48, & d^2_{43} + d^2_{31} &= 48 + 36, & d^3_{41} &= 48;\\
d^2_{42} &= 96, & d^2_{43} + d^2_{32} &= 48 + 60, & d^3_{42} &= 96.
\end{aligned}$$

$$\mathbf{D}^3 = \begin{pmatrix} 0 & 48 & 30 & 54 \\ 42 & 0 & 40 & 96 \\ 36 & 60 & 0 & 90 \\ 48 & 96 & 48 & 0 \end{pmatrix}$$

and

$$\mathbf{P} = \begin{pmatrix} 1 & 2 & 3 & 4 \\ 1 & 2 & 3 & 1 \\ 1 & 2 & 3 & 1 \\ 1 & 1 & 3 & 4 \end{pmatrix}.$$

The final step is to calculate $\mathbf{D}^4$ from $\mathbf{D}^3$:

$$
\begin{array}{lll}
d_{12}^3 = 48, & d_{14}^3 + d_{42}^3 = 54 + 96, & \text{set } d_{12}^4 = 48; \\
d_{13}^3 = 30, & d_{14}^3 + d_{43}^3 = 54 + 48, & d_{13}^4 = 30; \\
d_{21}^3 = 42, & d_{24}^3 + d_{41}^3 = 96 + 48, & d_{21}^4 = 42; \\
d_{23}^3 = 40, & d_{24}^3 + d_{43}^3 = 96 + 48, & d_{23}^4 = 40; \\
d_{31}^3 = 36, & d_{34}^3 + d_{41}^3 = 90 + 48, & d_{31}^4 = 36; \\
d_{32}^3 = 60, & d_{34}^3 + d_{42}^3 = 90 + 96, & d_{32}^4 = 60.
\end{array}
$$

Thus

$$
\mathbf{D}^4 = \begin{pmatrix} 0 & 48 & 30 & 54 \\ 42 & 0 & 40 & 96 \\ 36 & 60 & 0 & 90 \\ 48 & 96 & 48 & 0 \end{pmatrix}
$$

and

$$
\mathbf{P} = \begin{pmatrix} 1 & 2 & 3 & 4 \\ 1 & 2 & 3 & 1 \\ 1 & 2 & 3 & 1 \\ 1 & 1 & 3 & 4 \end{pmatrix}.
$$

The times of the shortest paths are given in $\mathbf{D}^4$ and the shortest paths themselves can be found by interpreting $\mathbf{P}$. As an example, to find the shortest path from town 3 to town 4 we see that $p_{34} = 1$. Thus the path begins 3–1. We now locate p_{14}, that is, the second town on the shortest path from town 1 to town 4. As $p_{14} = 4$, the complete path is 3–1–4. The other paths can be found in a similar manner.

The procedure illustrated above is now generalized.

Floyd's Method

Objective: To find the shortest path between each pair of vertices in a weighted graph G with n vertices and edge weight c_{ij}.

Steps

(1) Define $n \times n$ matrix $\mathbf{D} = (d_{ij})$, where d_{ij} is the length of the shortest path from vertex i to vertex j. Initially, set $d_{ij} = c_{ij}$ for all i and j. Define $n \times n$ matrix $\mathbf{P} = (p_{ij})$, where p_{ij} is the second vertex on the shortest path from vertex i to vertex j. Initially, set $p_{ij} = j$ for all j.

(2) For all k, $k = 1, 2, \ldots, n$, and for all i, $i = 1, 2, \ldots, n$, and for all j, $j = 1, 2, \ldots, n$, perform steps (3) and (4).

(3) Set d_{ij} to become the minimum of d_{ij} and $d_{ik} + d_{kj}$.

(4) Whenever $d_{ij} > d_{ik} + d_{kj}$ in step (3), set p_{ij} to become p_{ik}.

3.2.4. Exercises

1. Apply Dijkstra's method to find the shortest path from vertex 1 to vertex 10 in the following distance table:

	1	2	3	4	5	6	7	8	9
2	3								
3	4	6							
4	5	7	11						
5	—	9	12	9					
6	—	—	14	8	19				
7	—	—	15	12	20	27			
8	—	—	—	—	23	29	40		
9	—	—	—	—	26	30	41	50	
10	—	—	—	—	—	31	42	55	60

2. Find the shortest paths between all pairs of vertices 1, 2, 3, and 4 in the above table using Floyd's method.

*3. Compare the efficiency, in terms of the number of elementary arithmetical operations, in using Dijkstra's and Floyd's methods to find the shortest path between (a) a given pair of vertices in a weighted graph and (b) all pairs of vertices in a weighted graph.

3.3. The Maximum-Flow Problem

3.3.1. A Simple Numerical Example

The tourist area of Scenic Valley, introduced in Section 3.1, has recently been discovered to contain oil at Geyser Town (GT). A network of pipes has been built to transport the oil from GT (town 1), the source, to Twin Streams (town 5), the sink, which is the only port. The network is shown in Fig. 3.6. The capacity of the pipes to pump oil, in units of 10 barrels per hour, is also shown. The question is to decide what is the maximum amount of oil that can be pumped hourly from Geyser Town to Twin Streams.

It is assumed that each pipe can accept no more oil than its hourly capacity and that there is a very large quantity of oil at Geyser Town waiting to be pumped. We further assume that there is no loss of the oil at intermediate towns. This assumption is called the *conservation of flow*. This means that for each town other than the source (Geyser Town) and the sink (Twin Streams) the amount of oil flowing into the town equals the amount of oil flowing out of it.

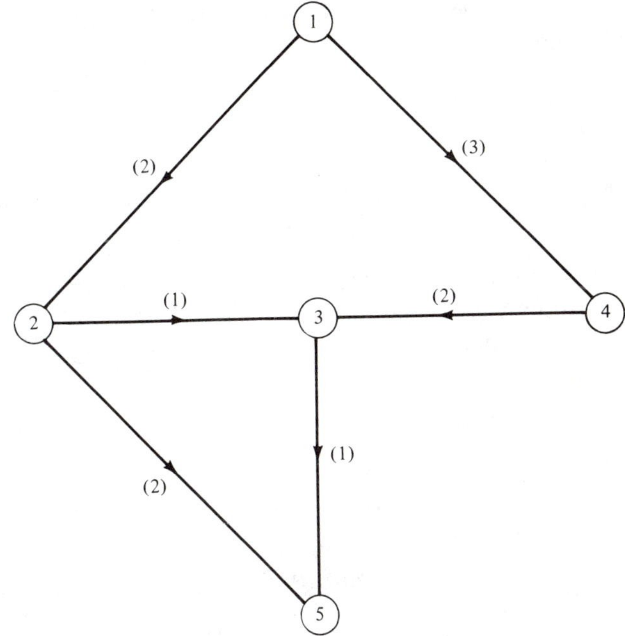

Figure 3.6. Network for Transporting Oil from Town 1 to Town 5.

We now build a mathematical model of the problem which will help us to decide how to assign hourly flows to the pipes in order to maximize flow. Let

c_{ij} = the hourly capacity of pipe (i, j) joining towns i and j;

f_{ij} = the amount of oil pumped in 10-barrel units directly from town i to town j;

F = the maximum amount of oil pumped hourly.

Let us begin with town 1. It is connected by pipes (1,2) and (1,4) to towns 2 and 4, respectively. The total amount of oil pumped from town 1 is then

$$f_{12} + f_{14}.$$

According to our definition, we can set this equal to F:

$$f_{12} + f_{14} = F.$$

This total quantity F must ultimately reach town 5 as we have assumed conservation of flow. The amount of oil reaching town 10 is

$$f_{25} + f_{35}.$$

Thus we can set

$$f_{25} + f_{35} = F.$$

We also have a constraint for each intermediate town based on the conservation of flow. Consider town 2. The amount of oil entering town 2 is f_{12}. The amount of oil leaving town 2 is

$$f_{23} + f_{25}.$$

Thus

$$f_{12} = f_{23} + f_{25}$$

or

$$f_{12} - f_{23} - f_{25} = 0.$$

We can build up a similar constraint for each other intermediate town. The last set of constraints are based on the capacity of each pipe, as defined in Fig. 3.6. We also assume that the amount that can be pumped is nonnegative. Thus we have

$$0 \le f_{ij} \le c_{ij} \qquad \text{for each pipe } (i, j).$$

The complete mathematical formulation is

$$\text{Maximize } F$$

subject to

$$\begin{aligned}
f_{12} + f_{14} \qquad\qquad &= F,\\
f_{12} \qquad - f_{23} - f_{25} &= 0,\\
f_{23} + f_{43} - f_{35} \qquad &= 0,\\
f_{14} - f_{43} \qquad\qquad &= 0,\\
f_{25} + f_{35} \qquad\qquad &= F,
\end{aligned}$$

$$\begin{matrix} 0 \le f_{12} \le 2, & 0 \le f_{23} \le 1, & 0 \le f_{14} \le 3,\\ 0 \le f_{43} \le 2, & 0 \le f_{25} \le 2, & 0 \le f_{35} \le 1. \end{matrix}$$

We develop a general formulation in the next section.

3.3.2. A General Formulation

Let n equal the number of nodes in the network and let the source and sink be denoted by network nodes p_1 and p_n, respectively. In its general form the maximum-flow problem is

$$\text{Maximize } F$$

subject to

$$\sum_{\substack{\text{all arcs}\\(p_i, p_j)}} f_{ij} - \sum_{\substack{\text{all arcs}\\(p_j, p_i)}} f_{ji} = \begin{cases} F & \text{for } i = 1,\\ 0 & \text{for } i \ne 1 \text{ or } n,\\ -F & \text{for } i = n,\end{cases}$$

$$0 \le f_{ij} \le c_{ij} \qquad \text{for all arcs } (p_i, p_j).$$

We now turn to developing a method for solving this problem.

3.3.3. Minimal Cuts

Look at Fig. 3.6. If arcs (2,5) and (3,5) were removed, town 1 would be cut off from town 10 and it would be impossible to ship any oil. A set of arcs whose removal cuts off the source from the sink is called a *cut*. The capacity of a cut is defined to be the sum of the capacities of the arcs in the cut. The cut of least capacity in the network is called the *minimal cut*. The capacity of the cut in question is thus

$$c_{25} + c_{35} = 2 + 1 = 3.$$

The following theorem is helpful in developing a method to solve the problem.

Theorem 3.1 (The max-Flow, min-Cut theorem). *The maximum amount of source to sink flow in any network is equal to the capacity of the minimal cut.*

We delay the proof of this theorem until the end of this section. The result allows us to recognize when we have found the maximum possible flow. The labeling method, which we now explain, progressively loads the network with more source to sink flow. Once the total equals the capacity of the minimal cut, an optimal solution is at hand.

3.3.4. The Labeling Method

The reader may have noticed that we defined f_{ij} in Section 3.3.1 so that it was theoretically possible to have flow from town j to town i along arc (p_i, p_j) in the backwards direction. Although this is usually impossible in reality for technical reasons, we find the definition a useful mathematical device. For each arc (p_i, p_j) we define an oppositely directed arc (p_j, p_i) with zero capacity. Flows in opposite directions in a single arc have their magnitudes subtracted one from the other to produce a single flow in the direction of the larger flow. To illustrate this, if we have $f_{14} = 3$ and $f_{41} = 2$, the result is a flow of 1 unit from p_1 to p_4. That is, an adjustment is made and $f_{14} = 1$ and $f_{41} = 0$. The *excess capacity* e_{ij} of an arc (p_i, p_j) when it has flow f_{ij} is defined to be the additional flow that it could accommodate up to its capacity c_{ij}. That is

$$e_{ij} = c_{ij} - f_{ij}.$$

This concept allows us to change our minds about previous assignments and modify them. For instance, suppose we have a flow $f_{23} = 0$ in arc (p_2, p_3) with capacity $c_{23} = 1$. Now $e_{23} = c_{23} - f_{23} = 1 - 0 = 1$. Imaginary arc (p_3, p_2) has flow $f_{32} = 0$ and capacity $c_{32} = 0$. Thus $e_{32} = 0$. Now suppose that we decide to assign a flow of $f_{23} = 1$ to (p_2, p_3). The excess capacity of (p_2, p_3) is reduced to $e_{23} = c_{23} - f_{23} = 1 - 1 = 0$. The excess capacity of arc

(p_3, p_2) is increased by the amount just assigned; that is, set $e_{32}=1$. Although it is impossible in practice for (p_3, p_2) to accommodate any flow, this mechanism allows us to assign *notionally* a flow of 1 to (p_3, p_2) since its excess capacity is 1. Suppose this is now done. It will allow us to change our minds and take back our original assignment of $f_{23}=1$. Defining $f_{32}=1$, the excess capacity of (p_3, p_2) is reduced to zero. The excess capacity of (p_2, p_3) is increased by 0 to 1 and the flows f_{23} and f_{32}, both 1, cancel each other out. We redefine $f_{23}=f_{32}=0$ and we are back where we started.

This process just described allows us to modify existing flows. Let us define it formally. Assume that arc (p_i, p_j) has flow f_{ij}. If a further flow of f'_{ij} is assigned to it we define

$$e_{ij} \quad \text{to become} \quad e_{ij} - f'_{ij}$$

and

$$e_{ji} \quad \text{to become} \quad e_{ji} + f'_{ij}.$$

Extra flow assigned to arc (p_i, p_j) with $e_{ij} > 0$ is termed *forward flow*. Extra flow assigned in the reverse direction (because $e_{ji} > 0$) is termed *backwards flow*.

It may be that there is no defined direction for the link between two points p_i and p_j. In this case two oppositely directed arcs (p_i, p_j) and (p_j, p_i) are defined, each with capacity equal to the original link capacity. The notion of subtracting f_{ij} and f_{ji}, one from the other, still applies.

We now use these ideas in explaining the labeling method by applying it to the example problem. We define a label (a_i, b_i) for each point p_i in the network other than p_1. Let

$b_i =$ the amount of extra flow which can be transported from p_1 to p_i, over and above the current flow assignment, while obeying arc capacity.

Set

$$b_1 = \infty.$$

The other points are progressively labeled with their b_i's as the method proceeds. Set

$$f_{ij} = 0$$

and

$$e_{ij} = c_{ij} \qquad \text{for each arc } (p_i, p_j).$$

At each iteration, any unlabeled point p_j directly connected by an arc (p_i, p_j), with excess capacity $e_{ij} > 0$, to a labeled point p_i is identified. For each identified point p_j, set

$$a_j = p_i$$

and

$$b_j = \text{Min}\{e_{ij}, b_i\}.$$

Thus we have labeled p_j with the label (a_j, b_j) indicating that it is possible to transport an extra flow of b_j units from p_1 to p_j via p_i.

We now perform this for the example problem. Please look at Fig. 3.6. Set

$$b_1 = \infty.$$

All other points are unlabeled. All f_{ij} values are zero, $e_{12} = 2$, and $e_{14} = 3$. We could label p_2 and p_4 with (1,2) and (1,3), respectively. Only one label is assigned at a time. The point with the smallest index is labeled first. Thus we label p_2. Next, p_3 can be labeled:

$$a_3 = 2$$

and

$$b_3 = \text{Min}\{b_2, e_{23}\} = \text{Min}\{2,1\} = 1.$$

Therefore $(a_3, b_3) = (2,1)$. In this way we can label p_5 as $(a_5, b_5) = (3,1)$.

Once the sink p_5, has been labeled, *breakthrough* has been achieved. We have now found, as displayed in the p_5 label, that we can transport a unit of flow from p_1 to p_5 via p_2 and p_3. The path of the assignment is deduced by looking at the a_i's: $a_5 = 3$, so the path has a final arc of (p_3, p_5). Also $a_3 = 2$, so the penultimate arc is (p_2, p_3). The complete path is $\langle 1,2,3,5 \rangle$. We now increase the flow in these arcs by $b_5 = 1$ unit:

$$f_{12} = f_{22} = f_{35} = 1.$$

The excess capacities of these arcs are reduced by the flow just assigned:

$$e_{12} = 2 - 1 = 1,$$
$$e_{23} = 1 - 1 = 0,$$
$$e_{35} = 1 - 1 = 0.$$

The excess capacities of the oppositely directed arcs are all increased by this flow:

$$e_{21} = 0 + 1 = 1,$$
$$e_{32} = 0 + 1 = 1,$$
$$e_{53} = 0 + 1 = 1.$$

We have just completed one iteration of the labeling method. All labels, except for b_1, are removed and we start over. The current flow assignment and the new excess capacities are shown in Fig. 3.7.

We begin the second iteration by labeling p_2 with $(a_2, b_2) = (1,1)$ as the excess capacity of (p_1, p_2) is now reduced to one. We cannot label p_3 as we did last time because $e_{23} = 0$. So we label p_5:

$$a_5 = 2$$

and

$$b_5 = \text{Min}\{b_2, e_{25}\} = \text{Min}\{1,2\} = 1.$$

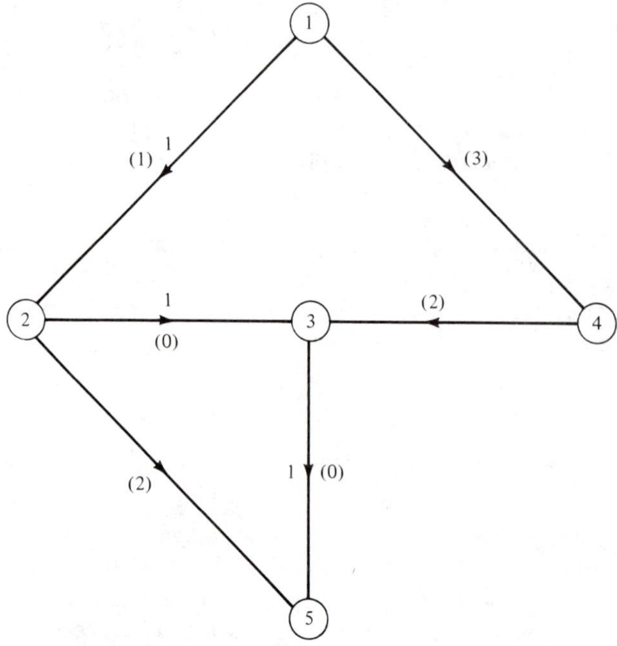

Figure 3.7. The Flow after the First Iteration.

Breakthrough has been achieved once more and we can augment the path $\langle a_2, a_5, 5\rangle = \langle 1,2,5\rangle$ with the flow $b_5 = 1$. Thus

$$f_{12} \quad \text{becomes} \quad f_{12}+1=1+1=2$$

and

$$f_{25} \quad \text{becomes} \quad 1.$$

We set

$$e_{25} \quad \text{to become} \quad e_{25}-1=2-1=2$$

and

$$e_{12} \quad \text{to become} \quad e_{12}-1=1-1=0.$$

The excess capacities of the oppositely directed arcs are increased by this flow:

$$e_{21} \quad \text{becomes} \quad e_{21}+1=1+1=2$$

and

$$e_{52} \quad \text{becomes} \quad e_{52}+1=0+1=1.$$

All labels are now removed. The current flow is shown in Fig. 3.8.

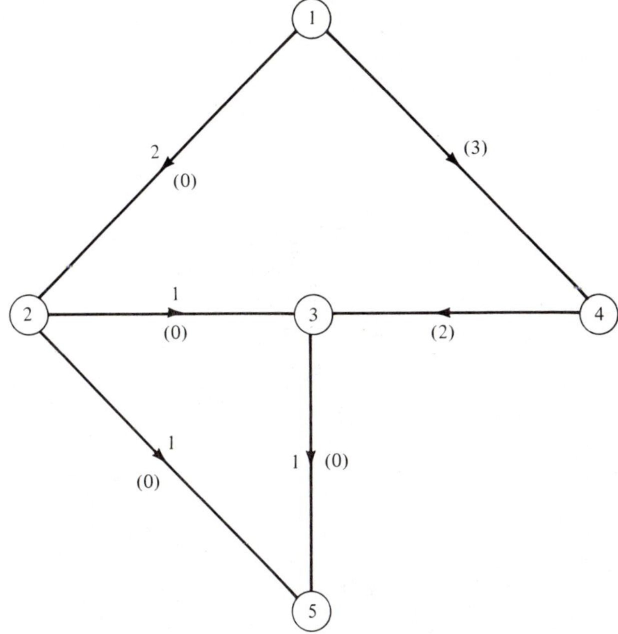

Figure 3.8. The Current Flow.

We begin the third iteration by labeling p_4 (We cannot label p_2 as $e_{12}=0$.): $a_4=1$ and $b_4=3$. We can now label p_3 with $(a_3,b_3)=(4,2)$. It now looks as if we have reached an impasse. However, we have the imaginary arc (p_3,p_2) with excess capacity $e_{32}=1>0$. We can use it to label p_2 with $(a_2,b_2)=(3,1)$. This allows us to label p_3 with $(a_5,b_5)=(2,1)$. We have found a final path $\langle 1,4,3,2,5\rangle$ to which we can add one further unit of flow:

$$f_{14}=1,$$
$$f_{43}=1,$$
$$f_{32}=1,$$
$$f_{25} \quad \text{becomes} \quad f_{25}+1=1+1=2.$$

The excess capacities are

$$e_{14}=3-1=2,$$
$$e_{43}=2-1=1,$$
$$e_{32} \quad \text{which becomes} \quad e_{32}-1=1-1=0,$$
$$e_{25} \quad \text{which becomes} \quad e_{25}-1=1-1=0.$$

The excess capacities of the oppositely directed arcs are increased by this flow:

$$e_{41}=0+1=1,$$
$$e_{34}=0+1=1,$$
$$e_{23}=0+1=1,$$
$$e_{52}=1+1=2.$$

Now an interesting thing occurs. We have a flow of $f_{23}=1$ in arc (p_2, p_3) and a flow of $f_{32}=1$ in arc (p_3, p_2). These flows are now canceled and we set $f_{23}=f_{32}=0$. We have removed our earlier flow of $f_{23}=1$. The final flows are shown in Fig. 3.9.

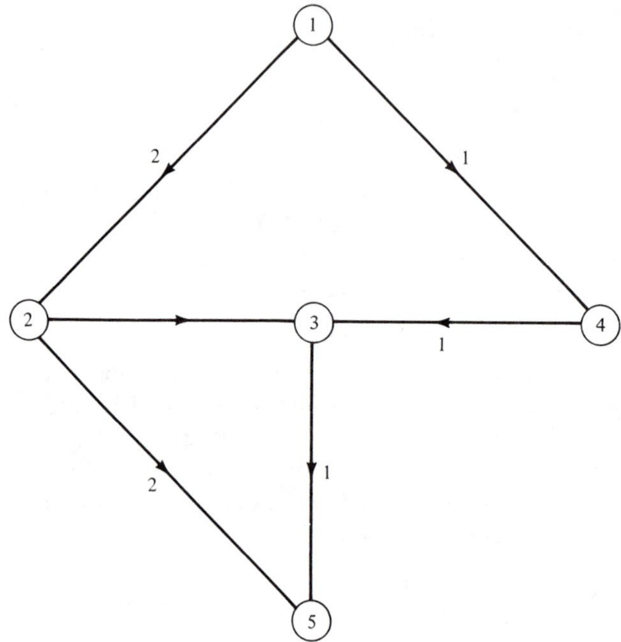

Figure 3.9. The Optimal Flow Assignment.

At the next iteration the labeling method fails to achieve breakthrough. The total flow $F=3$, which is the capacity of the minimal cut $\{(p_2, p_5), (p_3, p_5)\}$. The optimal solution has been found.

A formal statement of the labeling method is now given.

The Labeling Method

(1) Label the source p_1 with $(a_1, b_1)=(0, \infty)$.
(2) Set each f_{ij} value equal to that corresponding to a feasible flow, that is, one in which arc capacity and flow conservation are respected. In the

event of no other feasible flow assignments being available, set

$$\left.\begin{aligned} &f_{ij} = 0 \\ \text{and} \quad & \\ &e_{ij} = c_{ij} - f_{ij} \end{aligned}\right\} \text{for all arcs } (p_i, p_j) \text{ both real and introduced.}$$

(3) If there is no labeled point p_i directly connected to an unlabeled point p_j by an arc (p_i, p_j), terminate the method—the present flow assignment is optimal. Otherwise, continue.

(4) Find an arc (p_i, p_j) joining a labeled point p_i to an unlabeled point p_j. Set

$$a_j = p_i$$

and

$$b_j = \text{Min}\{e_{ij}, b_i\}.$$

(5) If the sink p_n is unlabeled go back to step (3). Otherwise go to step (6).

(6) For each arc (p_i, p_j) on the path of labeled points from p_1 to p_n, set

$$e_{ij} \quad \text{to become} \quad e_{ij} - b_n$$

and

$$f_{ij} \quad \text{to become} \quad f_{ij} + b_n.$$

For each arc (p_j, p_i) oppositely directed to arc (p_i, p_j) just identified, set

$$e_{ji} \quad \text{to become} \quad e_{ji} + b_n.$$

If, for any pair of arcs (p_i, p_j) and (p_j, p_i), it is true that f_{ij} and f_{ji} are both positive, set

$$\begin{aligned} f_{ij} &\quad \text{to become} \quad f_{ij} - |f_{ij} - f_{ji}|, \\ f_{ji} &\quad \text{to become} \quad f_{ji} - |f_{ij} - f_{ji}|, \\ e_{ij} &\quad \text{to become} \quad e_{ij} + |f_{ij} - f_{ji}|, \\ e_{ji} &\quad \text{to become} \quad e_{ji} + |f_{ij} - f_{ji}|. \end{aligned}$$

Erase all labels except that of p_1 and return to step (3).

Proof of Theorem 3.1

Let F equal the maximum possible source to sink flow found by the labeling method. The removal of the arcs in the minimal-cut set S creates two separate digraphs. Let the sets of vertices of these two digraphs be V and V'. We assume that p_1, the source, is in V and that p_n, the sink, is in V'.

Consider the arcs joining V and V'. The net sum of the flows between V and V' must be F for otherwise F would not be the maximum flow possible. That is

$$\sum_{\substack{p_i \in V \\ p_j \in V'}} f_{ij} - \sum_{\substack{p_k \in V' \\ p_m \in V}} f_{km} = F.$$

Further, for each arc (p_i, p_j) joining V to V', we must have

$$f_{ij} = c_{ij} \tag{3.1}$$

or else p_j could be labeled with forward flow from p_i. Summing (3.1) over all arcs in C:

$$\sum_{\substack{p_i \in V \\ p_j \in V'}} f_{ij} = \sum_{\substack{p_i \in V \\ p_j \in V'}} c_{ij}. \tag{3.2}$$

Similarly, for each arc (p_k, p_m) joining V' to V, we must have $f_{km} = 0$. Otherwise, vertex p_k could be labeled with backwards flow from p_m. Summing over all arcs connecting V' to V:

$$\sum_{\substack{p_i \in V \\ p_j \in V'}} f_{ij} - \sum_{\substack{p_k \in V' \\ p_m \in V}} f_{km} = C(S) - 0 = F,$$

where $C(S)$ equals the capacity of S.

Thus the maximum flow equals the capacity of the minimal-cut set. As the maximum flow must be less than or equal to the capacity of any cut set and we have equality, the result follows.

3.3.5. Exercises

1. Find the minimum cut in the network given in problem 1 of Section 3.2.4 where the entries now represent capacities. The source is node 1 and the sink is node 10.
2. Solve the maximum-flow problem posed in problem 1 above by the labeling method.

*3. Suppose two nations A and B are at war. A has the option of bombing B's territory, which contains a network of roads and bridges. Nation A wishes to disrupt the flow of materials in B's territory. Using the notions of this chapter explain why it is likely to be more advantageous for A to bomb the bridges rather than the roads.

3.4. The Minimum-Cost-Flow Problem

3.4.1. A Simple Numerical Example

Recall the problem of Section 3.3.1. Suppose now that the oil company operating in Scenic Valley has established unit operating costs for pumping the oil in the network shown in Fig. 3.6. The cost to pump one unit (10 barrels of oil) along each pipe, as well as its capacity, is shown in Fig. 3.10.

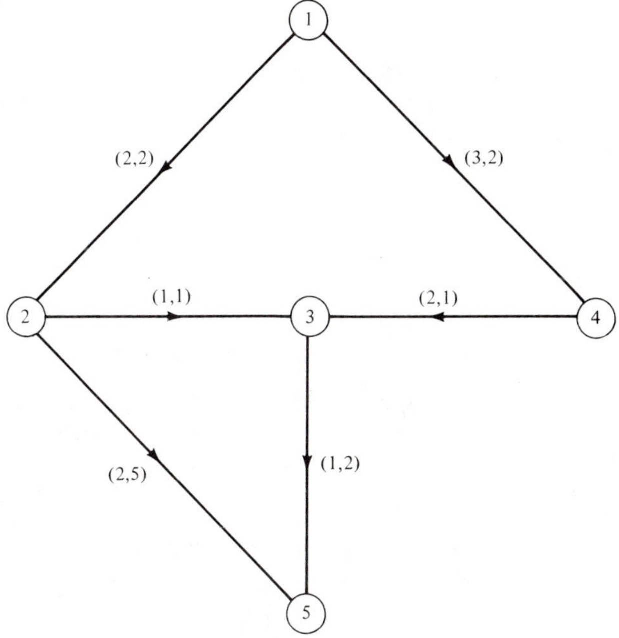

Figure 3.10. The Network for the Minimum-Cost-Flow Problem.

Each arc (i, j) in Fig. 3.10 has an ordered pair (c_{ij}, d_{ij}) associated with it, where c_{ij} is its capacity and d_{ij} is its unit cost. The company knows that a maximum of three units can be pumped hourly from source to sink. However, rather than run the system at capacity, it decides to pump only two units per hour. How can this be done at minimum total cost? The conservation of flow at intermediate points is once again assumed.

We can build a mathematical model of the problem, which is very similar to that of the maximum-flow problem of Section 3.3.1:

$$\text{Minimize} \sum_{\substack{\text{all arcs} \\ (p_i, p_j)}} d_{ij} f_{ij} \; (= Z)$$

subject to

$$f_{12} + f_{14} = 2,$$

$$f_{12} - f_{23} - f_{25} = 0,$$

$$f_{23} + f_{43} - f_{35} = 0,$$

$$f_{14} - f_{43} = 0,$$

$$f_{25} + f_{35} = 2,$$

$$0 \le f_{12} \le 2, \quad 0 \le f_{23} \le 1, \quad 0 \le f_{14} \le 3, \quad 0 \le f_{43} \le 2,$$

$$0 \le f_{25} \le 2, \quad \text{and} \quad 0 \le f_{35} \le 1.$$

The general formulation is given in the next section.

3.4.2. A General Formulation

Let n equal the number of nodes in the network. Let the source and the sink be denoted by network nodes p_1 and p_n, respectively. Let F be the given quantity to be transported at minimum cost. In its general form the minimum-cost-flow problem is

$$\text{Minimize} \sum_{\substack{\text{all arcs}\\(p_i,p_j)}} d_{ij} f_{ij}$$

subject to

$$\sum_{\substack{\text{all arcs}\\(p_i,p_j)}} f_{ij} - \sum_{\substack{\text{all arcs}\\(p_j,p_i)}} f_{ji} = \begin{cases} F & \text{for } i=1, \\ 0 & \text{for } i \neq 1 \quad \text{or} \quad n, \\ -F & \text{for } i=n, \end{cases}$$

$$0 \leq f_{ij} \leq c_{ij} \qquad \text{for all arcs } (p_i, p_j).$$

Naturally if F exceeds the capacity of the minimum cut, the problem has no feasible solution. If $F=1$ and

$$c_{ij}=1 \qquad \text{for all arcs } (p_i, p_j),$$

then the problem reduces to the shortest-path problem of Section 3.2.

3.4.3. A Minimum-Cost-Flow Method

We now explain a simple, iterative method to solve the problem. More efficient (but quite complicated) methods are referenced in the suggestions for further reading at the end of this book. It is assumed that the reader is familiar with Section 3.3. At each iteration the algorithm identifies among all the source to sink paths with excess capacity, the one with the least total cost. Flow equal to the excess capacity is then added to this path. The process continues in this way until the total source to sink flow equals F, the given quantity to be shipped. Of course the amount assigned in the last iteration may be strictly less than excess capacity in order to ensure that exactly F units (and not more) are shipped.

There is an important point which must be made about the way in which the total cost of each path is calculated. The cost of arcs oriented in the same direction as the source to sink path (forward flow) are added to the total cost. The cost of arcs oriented in the opposite direction to the source to sink path (backwards flow) are subtracted from the total cost. We now use the method to solve the problem of Section 3.4.1.

We begin by setting

$$f_{ij}=0 \qquad \text{for all arcs } (p_i, p_j).$$

A shortest path from p_1 to p_n in Fig. 3.10 is

$$\langle 1,2,3,5\rangle$$

with a cost of

$$d_{12}+d_{23}+d_{35}=2+1+2=5.$$

This path has excess capacity because, initially,

$$e_{ij}=c_{ij} \qquad \text{for all arcs } (p_i, p_j).$$

The excess capacity of this path is one unit. This amount is assigned to the path and the same bookkeeping as for the labeling method is performed:

$$f_{12}=f_{23}=f_{35}=1.$$

The excess capacities of these arcs are reduced by the flow just assigned:

$$e_{12}=2-1=1,$$
$$e_{23}=1-1=0,$$
$$e_{35}=1-1=0.$$

Arcs with zero excess capacity have their costs temporarily assigned to an arbitrarily large number which we denote by ∞. The excess capacities of the oppositely directed arcs are all increased by this flow:

$$e_{21}=0+1=1,$$
$$e_{32}=0+1=1,$$
$$e_{53}=0+1=1.$$

The unit cost of the newly created arcs are defined as

$$f_{21}=-2,$$
$$f_{32}=-1,$$
$$f_{53}=-2,$$

that is, the negative of the unit cost of the oppositely directed arc. The current flow assignment and the new arc labels are shown in Fig. 3.11. A shortest path from p_1 to p_n with excess capacity in Fig. 3.11 is

$$\langle 1,4,3,2,5\rangle$$

with a cost of

$$d_{14}+d_{43}+d_{32}+d_{25}=2+1-1+5=7.$$

This path has an excess capacity of one unit.

This amount is assigned to the path. When the bookkeeping is performed flow in arcs (2,3) and (3,2) are canceled. We have now assigned a total of two units. As this equals F, the method is terminated. The final flows are shown in Fig. 3.12.

The total cost can be calculated as follows for paths $\langle 1,2,5\rangle$ and $\langle 1,4,3,5\rangle$:

$$\langle 1,2,5\rangle : f_{12}d_{12}+f_{25}d_{25}=(1\times 2)+(1\times 5)=7$$

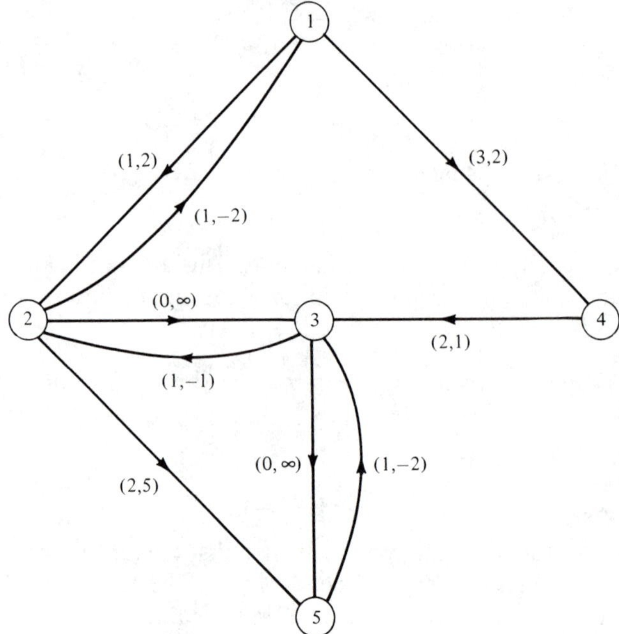

Figure 3.11. The Flow after the First Iteration.

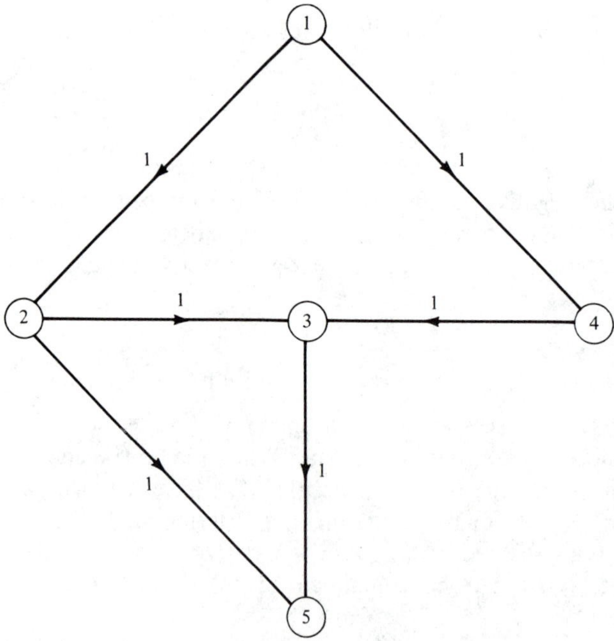

Figure 3.12. The Optimal Flow Assignment.

and

$$\langle 1,4,3,5\rangle : f_{14}d_{14} + f_{43}d_{43} + f_{35}d_{35} = (1\times 2)+(1\times 1)+(1\times 2) = 5.$$

$$Z \text{ (the total cost)} = 12.$$

A formal statement of the method is now given.

Summary of the Minimum-Cost-Flow Method

Steps

(1) Set

$$f_{ij} = 0$$

$$e_{ij} = c_{ij} \qquad \text{for all arcs } (p_i, p_j),$$

$$G = 0, \text{ denoting the } p_1 - p_n \text{ flow assigned so far.}$$

(2) Identify the $p_1 - p_n$ path, of least cost with positive excess capacity B. For each arc (p_i, p_j) on this path set

$$B = \text{Min}\{P, F - G\},$$

$$e_{ij} \quad \text{to become} \quad e_{ij} - B,$$

$$f_{ij} \quad \text{to become} \quad f_{ij} + B,$$

$$d_{ij} \quad \text{to become} \quad \infty \text{ if } e_{ij} = 0,$$

$$G \quad \text{to become} \quad G + B.$$

For each arc (p_j, p_i) oppositely directed to arc (p_i, p_j) just identified, set

$$e_{ji} \quad \text{to become} \quad e_{ji} + B$$

and

$$d_{ji} \quad \text{to become} \quad -d_{ij}.$$

If for any pair of arcs (p_i, p_j) and (p_j, p_i), $f_{ij} > 0$ and $f_{ji} > 0$ set

$$f_{ij} \quad \text{to become} \quad f_{ij} - |f_{ij} - f_{ji}|,$$

$$f_{ji} \quad \text{to become} \quad f_{ji} - |f_{ij} - f_{ji}|,$$

$$e_{ij} \quad \text{to become} \quad e_{ij} + |f_{ij} - f_{ji}|,$$

$$e_{ji} \quad \text{to become} \quad e_{ji} + |f_{ij} - f_{ji}|.$$

(3) If $G < F$, go back to step (2). If $G = F$, terminate. The current f_{ij} values indicate a least-cost solution.

3.4.4. Exercises

1. Consider problem 2 of Section 3.3.5. Consider this problem as a minimum-cost-flow problem, where each arc has unit cost equal to its capacity. Assume that the maximum quantity possible is to be shipped from node 1 to node 10. Solve this problem by the method of Section 3.4.3.

*2. Devise a method for transforming a minimum-cost-flow problem with multiple sources and multiple sinks into a problem with a single source and a single sink.

*3. Use the strategy developed in problem 2 above to solve the transportation problem of Section 1.2 as a minimum-cost-flow problem. Use the method of Section 3.4.3.

3.5. Activity Networks

3.5.1. A Simple Activity Networks Problem

The San Rosé Vineyards has an ambitious young manager, Pepe Santana, who has decided to advertise his company's wines in the following way. He plans to issue invitations to hundreds of selected people to come, a few at a time, to a free wine tasting. He has selected 20 combinations of six wines each. The invitation explains that a person will be allowed to taste any one combination of the wines described on the invitation. The person must return the invitation indicating his choice. Further, the tasting sessions will be run to a strict timetable and those invited must also indicate the time slot in which they wish to come.

Pepe plans to place each combination of wines on a numbered table and guide people to their choice of wines by handing them a numbered disk. The actual sampling time is only a small factor in the overall operation and is ignored. It is the preparation that needs analysis. There are many related activities involved in the running of the program. If he is to keep to his timetable, he must discover the relationships between activities and their duration times. To this end he compiles a list of the essential activities, their precedence activities, and their estimated duration times in minutes as shown in Table 3.5. The *precedence activities* for each activity are the activities which must be completed before the activity can commence. With this information he can draw a *precedence diagram*, shown in Fig. 3.13. Each point represents an activity and each arc represents a precedence relationship according to Table 3.5. The number of each activity is shown above its point in Fig. 3.13 and its duration time below it. There are two extra points: α, representing the fact that a new tasting cycle is about to begin, and ω, representing the fact that the actual tasting can now commence. We temporarily ignore the numbers inside the points.

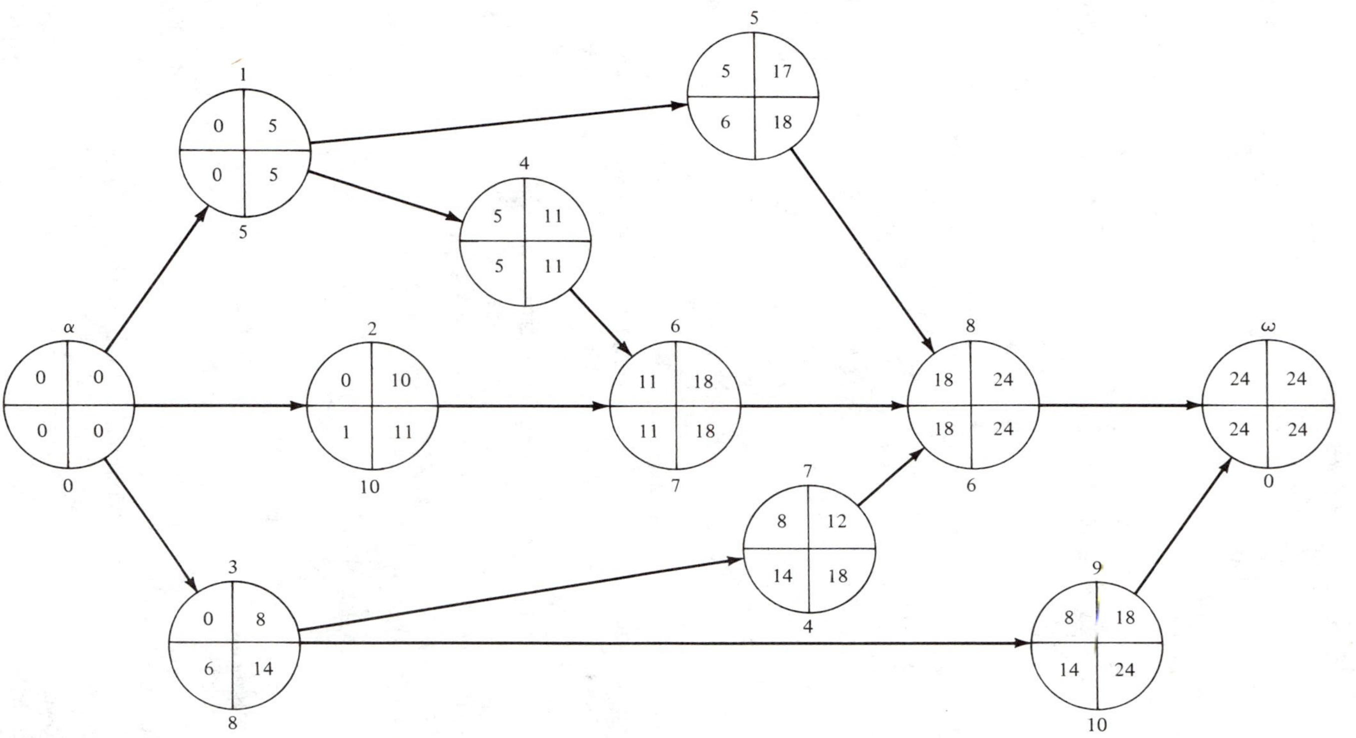

Figure 3.13. The Precedence Diagram for the Example Problem.

Table 3.5

Activity	Precedence	Duration time (min)
1. Clean up the tasting room	—	5
2. Locate desired wines in cellar	—	10
3. Welcome guests	—	8
4. Clean the glasses	1	6
5. Set up the tables and chairs	1	12
6. Pour the wines	2,4	7
7. Hand out seating numbers to guests	3	4
8. Set out the wines on the tables	5,6,7	6
9. Mingle with guests and then ask them into the tasting room	3	10

As we can see from Table 3.5, it is possible for some activities to be carried out simultaneously. Pepe assumes that he always has enough helpers to make this possible. The network is acyclic. If it were not, no activity on a cycle could ever be started. Some of the activities have a relatively short duration time. They can be fitted in anywhere in quite a long span of time without holding anything else up. On the other hand, some activities are *critical* in the sense that if they take longer than planned or if they are not done as soon as possible then the whole program will be delayed.

Pepe realizes that he must scrutinize the progress of the critical activities. If any critical activity appears to be falling behind schedule, he must divert extra resources to it at the expense of the noncritical activities to keep it on schedule. What Pepe wants to know is what is the *critical path* through the network from α to ω, that is, the path of critical activities. The time of this path (it is not necessarily unique) is the minimum possible duration time of the program as a whole. In the next section we explain a method that will produce this information.

3.5.2. The Critical Path Method

First, for each activity we must discover an *earliest start time*. This is the earliest time that the activity could be started given the fact that it may not be possible to begin it before other activities have been completed. We define the earliest starting time of the starting point α to be zero. Second, we must discover the *latest finish time* for each activity. This is the latest time by which the activity must be completed if the whole program is to finish at the earliest possible time (on schedule). Third, we can also define a *latest start time* for each activity. This is the latest time by which the activity must be started if the whole program is to finish on schedule. Finally, we define an *earliest finish time* for each activity. This is the earliest time the activity could be finished given its own duration time and the precedence relations.

Each critical activity will have its earliest start time equal to its latest start time. There will be no room for arbitrarily delaying the start of any critical activities if the program is to be completed on time. Noncritical activities will have latest start times which are strictly later than earliest start times. The difference between the two times we call the *float*.

The critical path method will produce a schedule that shows for each activity: its earliest and latest start times, its earliest and latest finish times, and its float. We can examine this information to identify the critical path. If things start to go wrong we can channel resources from noncritical activities knowing, by their floats, how much they can be safely delayed.

We now show how to calculate these times for each activity. For each activity i, let

$$es_i = \text{its earliest start time},$$

$$lf_i = \text{its latest finish time},$$

$$ls_i = \text{its latest start time},$$

$$ef_i = \text{its earliest finish time},$$

$$t_i = \text{its duration}.$$

We define

$$t_\alpha = t_\omega = 0.$$

We divide the point representing activity i in the precedence diagram into the quadrants as shown in Fig. 3.14.

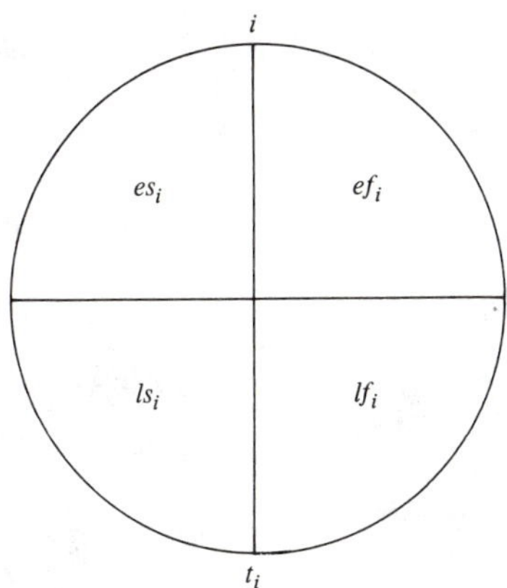

Figure 3.14. The Location of the Times.

The times ef_i and es_i are calculated using the following rules. Assume a network $N = (P, A)$ with point set P and arc set $A \subseteq P \times P$.

Rules

(1) Define $es_\alpha = 0$.
(2) Define $ef_i = es_i + t_i$.
(3) Define

$$es_i = \underset{\substack{j \\ (i,j)\in A}}{\text{Max}} \{ef_j\}.$$

Let us explain these rules:

(1) We assume that the project starts at time zero.
(2) As activity i takes time t_i, if its earliest start time is es_i then its earliest finish time is as shown.
(3) Activity i cannot start until all the activities which directly precede it are finished. Each activity j which directly precedes it will have the property that there is an arc (j, i) in the precedence diagram. Thus the latest among the earliest finishing times of these activities is the earliest that i can start.

Using rules (1)–(3) we can now calculate the earliest times ef_i and es_i for each activity i:

(1) $es_\alpha = 0$.
(2) $ef_\alpha = es_\alpha + t_\alpha = 0 + 0 = 0$.
(3) $es_1 = \text{Max}_\alpha\{ef_\alpha\} = 0$. Similarly, $es_2 = 0$ and $es_3 = 0$.
(2) $ef_1 = es_1 + t_1 = 0 + 5 = 5$.
$ef_2 = es_2 + t_2 = 0 + 10 = 10$.
$ef_3 = es_3 + t_3 = 0 + 8 = 8$.
(3) $es_5 = \text{Max}_1\{ef_1\} = 5$.
$es_4 = \text{Max}_1\{ef_1\} = 5$.
$es_7 = \text{Max}_3\{ef_3\} = 8$.
$es_9 = \text{Max}_3\{ef_3\} = 8$.
(2) $ef_5 = es_5 + t_5 = 5 + 12 = 17$.
$ef_4 = es_4 + t_4 = 5 + 6 = 11$.
$ef_7 = es_7 + t_7 = 8 + 4 = 12$.
$ef_9 = es_9 + t_9 = 8 + 10 = 18$.
(3) $es_6 = \text{Max}_{2,4}\{ef_2, ef_4\} = 11$.
(2) $ef_6 = es_6 + t_6 = 11 + 7 = 18$.
(3) $es_8 = \text{Max}_{5,6,7}\{ef_5, ef_6, ef_7\} = 18$.
(2) $ef_8 = es_8 + t_8 = 18 + 6 = 24$.
(3) $es_\omega = \max_{8,9}\{ef_8, ef_9\} = 24$.
(2) $ef_\omega = es_\omega + t_\omega = 24 + 0 = 24$.

We have now calculated the earliest times for all activities. They are shown in the appropriate quadrants in Fig. 3.13. As $ef_\omega = 24$, the earliest the program can be completed is in 24 minutes.

We now state rules for calculating the latest times. They require the value of ef_ω:

Rules

(4) Define $lf_\omega = ef_\omega$.
(5) Define $ls_i = lf_i - t_i$.
(6) Define

$$lf_i = \operatorname*{Min}_{\substack{j \\ (i,j) \in A}} \{ls_j\}.$$

Let us explain these rules:

(4) We assume that the program must finish as early as possible.
(5) As activity i takes time t_i, its latest finish time is equal to its latest start time plus its duration time; that is, $lf_i = ls_i + t_i$.
(6) No activity can be allowed to end after the latest time by which any successor must be begun.

Using rules (4)–(6) we can calculate the latest times ls_i and lf_i for each activity i:

(4) $lf_\omega = ef_\omega = 24$.
(5) $ls_\omega = lf_\omega - t_\omega = 24 - 0 = 24$.
(6) $lf_8 = \text{Min}_\omega\{ls_\omega\} = 24$.
$lf_9 = \text{Min}_\omega\{ls_\omega\} = 24$.
(5) $ls_8 = lf_8 - t_8 = 24 - 6 = 18$.
$ls_9 = lf_9 - t_9 = 24 - 10 = 14$.
(6) $lf_5 = \text{Min}_8\{ls_8\} = 18$.
$lf_6 = \text{Min}_8\{ls_8\} = 18$.
$lf_7 = \text{Min}_8\{ls_8\} = 18$.
(5) $ls_5 = lf_5 - t_5 = 18 - 12 = 6$.
$ls_6 = lf_6 - t_6 = 18 - 7 = 11$.
$ls_7 = lf_7 - t_7 = 18 - 4 = 14$.
(6) $lf_4 = \text{Min}_5\{ls_6\} = 11$.
$lf_3 = \text{Min}_{7,9}\{ls_7, ls_5\} = 14$.
$lf_2 = \text{Min}_6\{ls_6\} = 11$.
(5) $ls_4 = lf_4 - t_4 = 11 - 6 = 5$.
$ls_3 = lf_3 - t_3 = 14 - 8 = 6$.
$ls_2 = lf_2 - t_2 = 11 - 10 = 1$.
(6) $lf_1 = \text{Min}_{4,5}\{ls_4, ls_5\} = 5$.
(5) $ls_1 = lf_1 - t_1 = 5 - 5 = 0$.
(6) $lf_\alpha = \text{Min}_{1,2,3}\{ls_1, ls_2, ls_3\} = 0$.
(5) $ls_\alpha = lf_\alpha - t_\alpha = 0 - 0 = 0$.

We have now calculated the latest times for all activities. They are shown in the appropriate quadrants in Fig. 3.13. We note that $lf_\alpha = ls_\alpha = 0$. This must always happen if the calculations have been carried out without error. It provides a useful check.

We are now in a position to find the critical activities. Any activity i for which

$$es_i = ls_i$$

and

$$ef_i = lf_i$$

is critical.

The critical activities will form at least one path from α to ω. Each such path is termed a critical path. From our definitions, α and ω will always be critical. In our example there is only one critical path: $\langle \alpha, 1, 4, 6, 8, \omega \rangle$. We have found that the vital sequence of activities is:

- Activity 1—Clean up the tasting room.
- Activity 4—Clean the glasses.
- Activity 6—Pour the wines.
- Activity 8—Set out the wines on the tables.

If the completion of any of these activities is delayed, the whole project will be delayed.

We can now calculate the *float* for each activity—the amount of leeway we have in starting the activity without delaying the program completion. *Total float* tf_i, for each activity i is the difference between the maximum time available to complete the activity (which is $lf_i - es_i$) and its duration time t_i. That is,

$$tf_i = lf_i - es_i - t_i.$$

But by Rule (5),

$$ls_i = lf_i - t_i$$

and

$$tf_i = lf_i - t_i - es_i.$$
$$\therefore tf_i = ls_j - es_i.$$

Also by Rule (2):

$$ef_i = es_i + t_i$$

and

$$tf_i = lf_i - es_i - t_i$$
$$= lf_i - (es_i + t_i). \tag{3.3}$$
$$\therefore tf_i = lf_i - ef_i.$$

Using Eq. (3.3) we can calculate the total floats:

i:	α	1	2	3	4	5	6	7	8	9	ω
tf_i:	0	0	1	6	0	1	0	6	0	6	0.

Naturally the critical activities have zero total float.

As an example of the meaning of these computations, let us consider activity 7: hand out seating numbers to guests. The earliest time that this process can be started is $es_7 = 8$. The latest finish time is $lf_7 = 18$. Thus we have $lf_7 - es_7 = 18 - 8 = 10$ min in which to get it done. However, it takes only 4 min. Thus we have a float of $(lf_7 - es_7) - t_7 = (18 - 8) - 4 = 6$ min.

There is another kind of float which is useful in planning called *free float*. We shall explain its motivation by using activity 3. Activity 3 has an earliest finish time of $ef_3 = 8$. From Fig. 3.13 we can see that activities 7 and 9 directly follow activity 3. Consider activity 9. It has an earliest start time of $es_9 = 8$. Pepe wants to get each activity finished as soon as possible in case there are catastrophes near the end of the program. Thus he wants to start activity 9 as early as possible, namely, at minute 8. Thus activity 1 has a free float of no more than $es_9 - ef_3 = 8 - 8 = 0$. There is also activity 7 to consider. It has an earliest start time of $es_7 = 8$. It also confirms the fact that activity 1 has zero free float.

We define the free float ff_i of activity i to be

$$ff_i = \underset{\substack{j \\ (i,j) \in A}}{\text{Min}} \left\{ es_j - ef_i \right\}. \tag{3.4}$$

This definition does not cover ff_ω which we define as

$$ff_\omega = 0.$$

The reader should verify that the free float for any activity will never exceed its total float. Using Eq. (3.4) we can now calculate the free floats:

i:	α	1	2	3	4	5	6	7	8	9	ω
ff_i:	0	0	1	0	0	1	0	6	0	6	0.

This table tells Pepe how much real time he has available for each activity if he wishes to complete it at the earliest possible instant.

3.5.3. The Activity-Arc Model

We have been representing activities by network nodes and precedence relations by arcs. The more common approach to critical path analysis is to represent activities by arcs and the event that each is complete by a node. Naturally the two different approaches usually produce very different networks for any program. However, it is possible to construct one network from the other. There are advantages and disadvantages to each approach. The *activity-arc approach* usually requires the introduction of dummy or fictitious activities in order to produce the correct precedence relations. However, in the *activity-node approach* of the last two sections one has to introduce points α and ω. This approach allows one both to make changes to the precedence diagram and calculate the times much more easily.

3.5.4. Program Evaluation and Review Technique

We have not yet discussed the variation of the duration time of the activities. We have simply assumed each to be a known constant. This is seldom the case in realistic industrial projects or indeed in any real-life program. The exact times are unknown and can be estimated only by using some statistical process. If such is the case the Program Evaluation and Review Technique (PERT) is an appropriate method to use.

3.5.5. Exercises

1. Construct an activity-node network for the following program. Identify the critical path and calculate activity floats.

Activity	Precedence	Duration time (min)
1	11	19
2	17	11
3	12	20
4	11,15	16
5	12	3
6	7	2
7	7	5
8	8,12	4
9	8	7
10	7	8
11	—	1
12	4,5	1
13	9,10,3	6
14	9,10,3	4
15	1,2	9
16	16	97
17	14,16,13	3

2. Repeat problem 1 above for the following data:

Activity	Precedence	Duration time (min)
1	6,8	4
2	7	12
3	4,5	10
4	3	3
5	2	4
6	2	6
7	2	5
8	—	4
9	—	6

3. Consider problem 2 above. Suppose that $t_4 = 2$. How does this affect the calculations?

4. Consider the project of redecorating the interior of a house including stripping the old wallpaper, painting the walls, and laying down carpet. Assume that there are two people available. Devise a precedence diagram with about 15 activities. Apply the activity-node critical path method to it.

5. Carry out the critical path method on the following projects: changing a car tire; planning and staging a dinner party; and building a shed.

*6. Devise a method for finding the smallest number of people required to complete a program in the shortest possible time, assuming that the number of people needed for each activity is known.

PART II

APPLICATIONS

CHAPTER 4

Some Applications of Combinatorial Optimization Techniques

4.1. Facilities Layout

4.1.1. A Simple Numerical Problem

The Westminster Company has a problem. Their office space is too small and they are planning a new one-floor office building. The problem is how to answer the architect's question: How do you want the various facilities to be laid out? We begin by listing and enumerating the facilities:

(1) President's office.
(2) Space for President's secretary/receptionist.
(3) Typing pool.
(4) Sales Manager's office.
(5) Production Manager's office.
(6) Distribution Manager's office.
(7) Area for clerks and telephone operator.
(8) Boardroom.
(9) Washroom/Cafeteria.
(10) Storeroom.
(11) Building grounds.

Next a table is constructed whose entries represent the number of trips made by people between the various facilities. This table is shown as Table 4.1. The question we address is how to lay out the various facilities on the floor of the office block so as to minimize the total amount of walking done by everyone involved. The company makes the assumption that if two

Table 4.1

		Facilities										
		1	2	3	4	5	6	7	8	9	10	11
	2	8										
	3	5	25									
	4	4	11	10								
	5	6	14	11	3							
Facilities	6	7	9	12	4	4						
	7	2	30	44	5	6	5					
	8	2	3	0	2	2	2	0				
	9	2	3	75	2	2	2	39	2			
	10	0	2	14	0	0	0	15	0	2		
	11	82	2	22	2	2	2	20	0	31	64	

facilities are sited adjacently then the distance traveled in walking between them will be reduced drastically. So the question becomes one of which pairs of facilities should be sited adjacently in order to maximize the total number of walking trips saved.

We come now to devising a system for describing the adjacency structure of the layout. Consider Fig. 4.1. Here we have a possible layout of the building, where the areas are numbered according to the previous list, including the exterior region—the grounds in which the building is set.

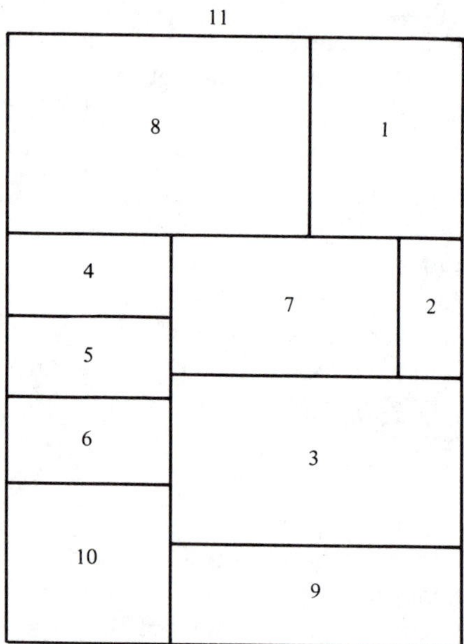

Figure 4.1. A Possible Office Layout.

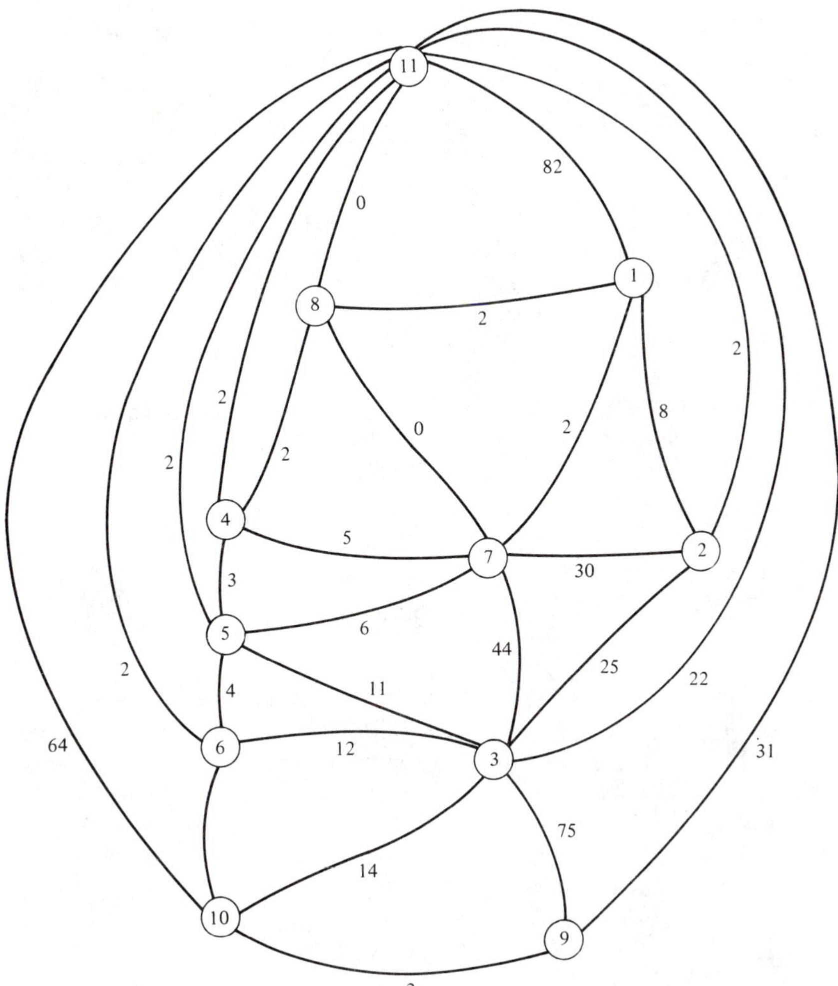

Figure 4.2. The Adjacency Graph for the Layout.

Figure 4.2 indicates which pairs of facilities are adjacent by using graph theory (see Chapter 5, Section 5.2 for an explanation of the terms used here). Each facility is represented by a vertex, and two vertices are joined by an edge if and only if their corresponding facilities have a positive length of wall in common.

Because the layout is to take place on one floor only, the corresponding adjacency graph is planar. Indeed because the layout does not have any facilities nested inside others the graph is maximally planar. If it was desirable to nest one facility inside another we could consider the two to be one superfacility for planning purposes. So for the rest of this chapter we assume that no nesting occurs and that adjacency graphs of layouts will be maximally planar.

We can associate with each edge in the graph in Fig. 4.2 a weight as shown. The weight for edge ij, joining vertices i and j, is the $i-j$ entry in Table 4.1. That is, the number of trips to be made between each pair of adjacent facilities is associated with the edge joining their vertices in the adjacency graph. The sum of these weights in Fig. 4.2 is 452. This is the total number of trips saved by locating certain pairs of facilities adjacently. All the trips in Table 4.1 still have to be made, but 452 of the trips are now of very short distance due to the facility adjacency.

The problem boils down to finding which pairs of facilities to locate adjacently in order to maximize the total number of trips saved. The layout in Fig. 4.2 saves 452 trips out of a total of 576. Can we possibly save more with a better layout? In the next few sections we develop a general solution technique for this problem.

4.1.2. The Problem in Graph Theoretic Terms

The problem described in the previous section can be modeled in terms of graph theory as follows. A complete graph G is formed, where each vertex represents a facility and each edge ij represents the possibility of locating facilities i and j adjacently. Each edge ij is given a weight equal to the number of trips made between facilities i and j. Any planar subgraph of G represents the adjacency structure of a layout for the facilities in a planar site. Because weights represent trips saved, it is logical to maximize the weight of the planar subgraph and so we wish to find a maximum weight planar subgraph of a complete weighted graph. We now develop a solution method for this problem.

There is the possibility of enumerating all possible solutions and choosing the best. We could begin to do this by recognizing that there are $\frac{1}{2}n(n-1)$ edges in any complete graph with n vertices. Any maximal planar subgraph of n vertices has $3n-6$ edges. So we could identify all combinations of $3n-6$ edges out of the existing $\frac{1}{2}n(n-1)$ edges. First, for a realistic size of n there is a large number of combinations. Second, we still have to check each combination to see whether or not it corresponds to a planar graph, that is, is feasible. Unfortunately checking for planarity requires a great deal of computational effort. We have to do better than complete enumeration. Unfortunately, the problem is NP-Complete (see Chapter 2). Loosely speaking, a problem is NP-Complete if it belongs to a special class of problems. It is unlikely that a polynomially time-bounded algorithm exists for any problem in the class. Hence we develop a heuristic method in the next section.

4.1.3. The Deltahedron Method

It is easy to show that the faces or regions of a maximal planar graph are all triangles—including the exterior face. So a maximal planar graph is equivalent to a polyhedron, all of whose faces are triangles. Such a polyhedron is sometimes called a *deltahedron*. Hence the name of the method.

The heuristic solution procedure builds up a maximal planar graph one vertex at a time. It begins by filling out the other half of the trip table. This is shown, for our numerical example, in Table 4.2. The columns have been

Table 4.2

		Facilities										
		1	2	3	4	5	6	7	8	9	10	11
Facilities	1	—	8	5	4	6	7	2	2	2	0	82
	2	8	—	25	11	14	9	30	3	3	2	2
	3	5	25	—	10	11	12	44	0	75	14	22
	4	4	11	10	—	3	4	5	2	2	0	2
	5	6	14	11	3	—	4	6	2	2	0	2
	6	7	9	12	4	4	—	5	2	2	0	2
	7	2	30	44	5	6	5	—	0	39	15	20
	8	2	3	0	2	2	2	0	—	2	0	0
	9	2	3	75	2	2	2	39	2	—	2	31
	10	0	2	14	0	0	0	15	0	2	—	64
	11	82	2	22	2	2	2	20	0	31	64	—
Column sums		118	107	218	43	50	47	166	13	160	97	227

Table 4.3

		Facilities										
		11	3	7	9	1	2	10	6	5	4	8
Facilities	1	82	5	2	2	—	8	0	7	6	4	2
	2	2	25	30	3	8	—	2	9	14	11	3
	3	22	—	44	75	5	25	14	12	11	10	0
	4	2	10	5	2	4	11	0	4	3	—	2
	5	2	11	6	2	6	14	0	4	—	3	2
	6	2	12	5	2	7	9	0	—	4	4	2
	7	20	44	—	39	2	30	15	5	6	5	0
	8	0	0	0	2	2	3	0	2	2	2	—
	9	31	75	39	—	2	3	2	2	2	2	2
	10	64	14	15	2	0	2	—	0	0	0	0
	11	—	22	20	31	82	2	64	2	2	2	0
Column sums		227	218	166	160	118	107	97	47	50	43	13

summed and reordered with the largest first down to the smallest, in Table 4.3. The first four vertices, corresponding to the first four columns in Table 4.3, are all mutually joined by edges to form a maximal planar graph on four vertices—a tetrahedron. It is shown in Fig. 4.3. This construction is motivated by the fact that perhaps we should locate the four most popular facilities all mutually adjacent.

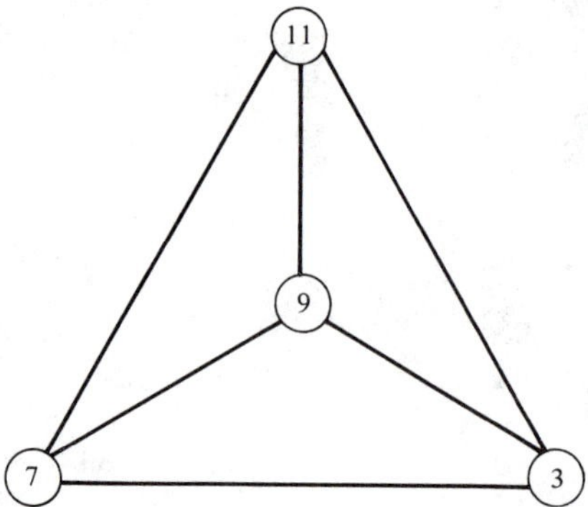

Figure 4.3. The Initial Tetrahedron.

The next vertex to be considered is the one corresponding to the next column in the order given in Table 4.3. It is vertex 1. It is inserted into a triangle of the maximal planar graph built up so far. A vertex, u say, is inserted into a triangle by the method illustrated in Fig. 4.4, that is, by joining it up by the edges to the vertices of the triangle.

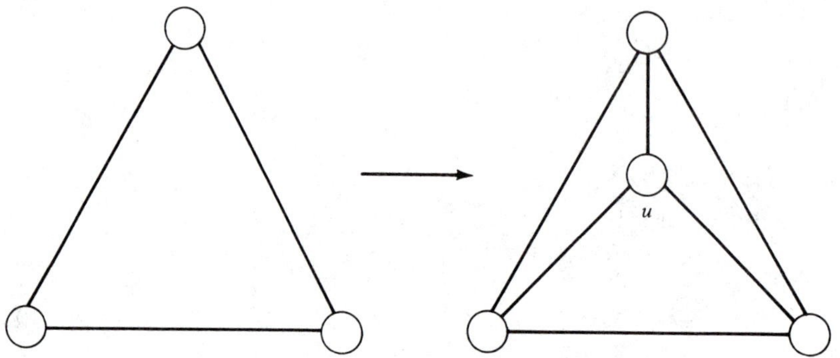

Figure 4.4. Vertex Insertion.

We have a choice of four triangles into which to insert vertex 1. Three of the triangles are self-evident in Fig. 4.3 and the fourth is the region exterior to the graph which is bound by edges: (7,11), (3,11), and (3,7). The four possible insertions are shown in Fig. 4.5.

The *weight* of an insertion is defined to be the sum of the weights of its three edges. The insertion with the largest weight is Fig. 4.5(b). Note that

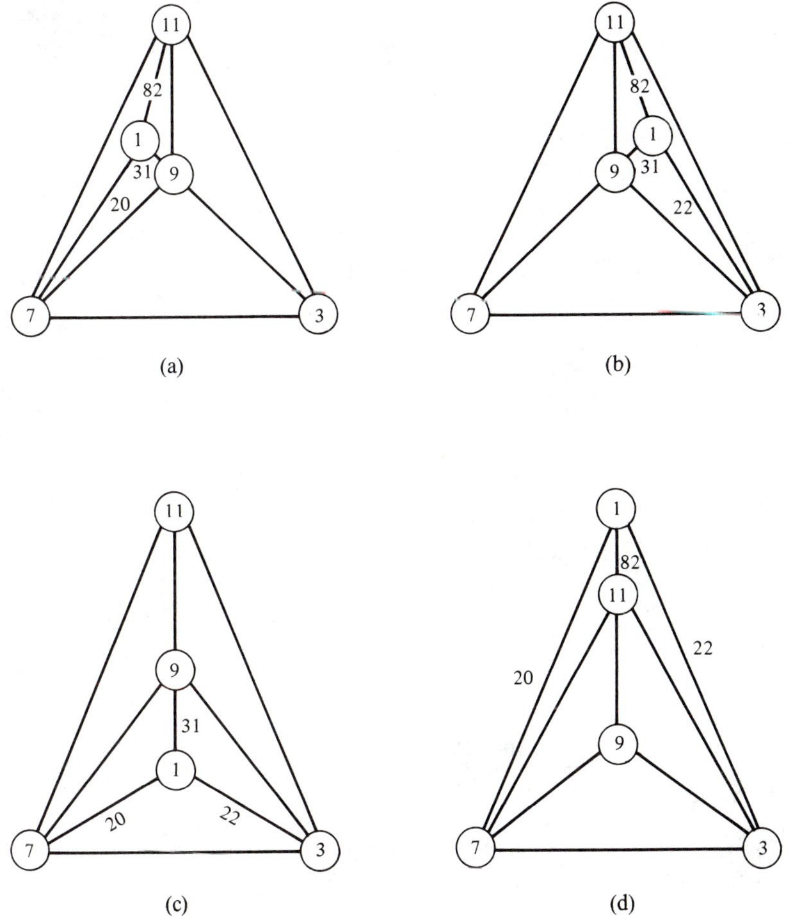

Figure 4.5. The Four Possible Insertions.

the weights of insertions, Figs. 4.5(a)–4.5(d) are $82+31+20$, $82+22+31$, $20+31+22$, and $82+22+20$, respectively.

The largest weight insertion is made a permanent part of the solution. So the maximal planar graph built up so far is given in Fig. 4.5(b).

The next vertex (number 2) in the order given in Table 4.3 is chosen. It is temporarily inserted in all the triangles of the graph in Fig. 4.5(b). The insertion with the largest weight is the insertion of vertex 2 into the triangle with vertices: 3, 7, and 9. The resulting maximal graph is shown in Fig. 4.6. This process is continued, inserting the vertices into the graph in Table 4.3, namely, 10, 6, 5, 4, and 8. The reader should verify that one possible sequence of graphs that can be constructed is that given in Fig. 4.7.

The final graph in Figure 4.7 (on page 169) is a prescription of which pairs of facilities are to be adjacent in the layout. Its total weight is 557. The layout corresponding to this is given in Fig. 4.8 (on page 170).

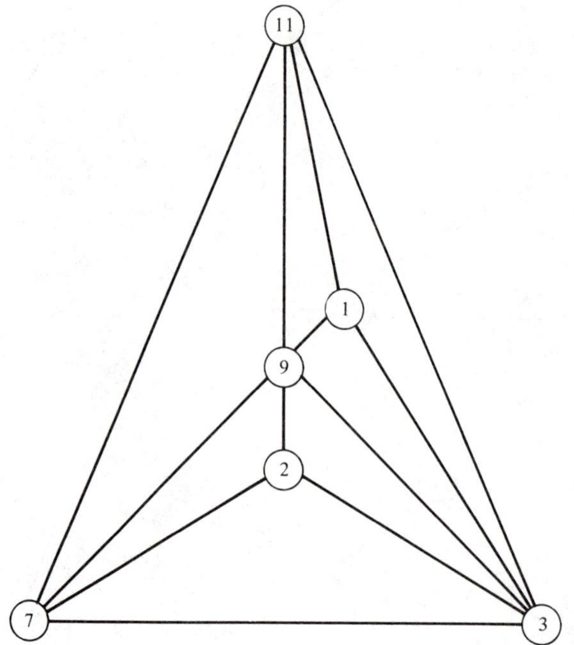

Figure 4.6. The Adjacency Graph with the First Six Facilities.

4.1.4. Exercises

1. Use the deltahedron method to solve the following layout problem where 1 represents the exterior region:

	1	2	3	4	5	6	7	8	9	10	11	12
2	7											
3	8	45										
4	11	92	31									
5	4	8	70	15								
6	9	0	93	17	17							
7	12	16	12	99	2	12						
8	10	77	4	8	9	3	17					
9	0	1	5	3	8	23	20	3				
10	40	14	12	6	88	39	51	1	41			
11	7	22	7	41	70	88	53	9	20	11		
12	17	19	11	20	72	81	12	8	85	50	21	
13	23	11	51	67	49	61	7	7	14	94	3	7

*2. Devise a procedure which takes the final adjacency graph and draws a scale plan of its layout, taking the areas of the facilities into account.

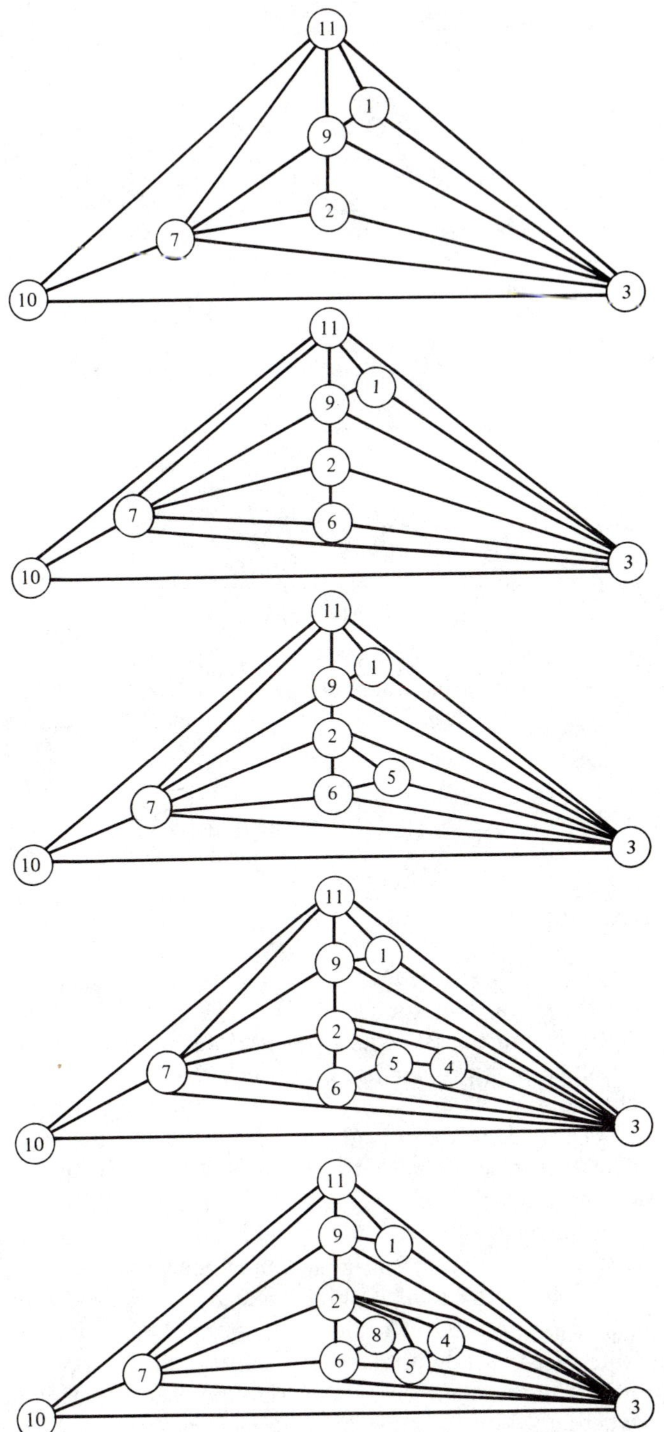

Figure 4.7. The Sequence of Adjacency Graphs Derived by the Deltrahedron Method.

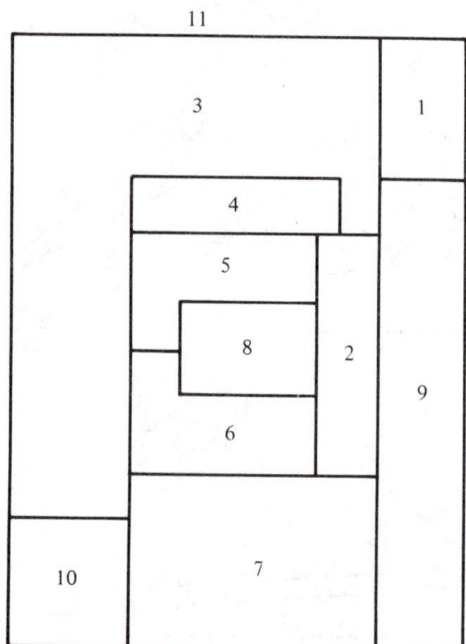

Figure 4.8. The Final Layout.

*3. Devise a modification of the deltahedron heuristic which lays out all the floors of an office block where there is interaction between facilities on different floors.

4.2. The Traveling-Salesman Problem

4.2.1. Introduction

The SV Tourist Agency, introduced in Section 3.1, has launched a promotion to boost the number of tourists of Scenic Valley. They are about to offer a package tour to tourists who fly in and out of Geyser Town (GT), which is the only airport with connections to towns outside the valley. They now offer a bus tour of the valley, visiting each town once, starting from and returning to GT. In order to estimate the costs of the tour the agency needs to know the least number of miles the bus must travel in making the tour.

The table of intertown distances is given in Table 3.1. The question is to find the closed sequence of towns (in which each is visited exactly once) of shortest distance. We can represent the towns by the vertices of a graph (see Chapter 5, Section 2), the roads joining them by edges, and weight each edge by the appropriate entry in Table 3.1. In graph theoretic terms we wish to find the least-weight Hamiltonian cycle of the graph.

This problem is an example of the well-known "Traveling-Salesman Problem," where a salesman rather than a bus is to tour a set of towns in least distance. If the tour did not have to return to the starting point, we have the shortest Hamiltonian path problem analyzed in Chapter 0. We can formulate the problem as an integer-programming problem as follows. Let

$$x_{ij}\begin{cases} =1 & \text{if the bus travels directly from town } i \text{ to town } j, \\ =0 & \text{otherwise;} \end{cases}$$

and

$$d_{ij} = \text{the distance from town } i \text{ to town } j.$$

Then the problem is to find a cycle which visits each town exactly once and which

$$\text{Minimizes } \sum_{i=1}^{10} \sum_{j=1}^{10} d_{ij} x_{ij}$$

subject to

$$\sum_{i=1}^{10} x_{ij} = 1, \qquad j = 1, 2, \ldots, 10,$$

and

$$\sum_{j=1}^{10} x_{ij} = 1, \qquad i = 1, 2, \ldots, 10.$$

Because the general problem of this type is NP-Complete (see Chapter 2) it is unlikely that realistically sized problems can be solved optimally in reasonable computing time. However, we do note in passing that surprisingly large problems have yielded optimal solutions to the application of algorithms. However we confine our attention to heuristics.

4.2.2. The Nearest-Neighbor Heuristic

We start with the obvious heuristic of beginning at a town and traveling to the nearest town not yet visited. Beginning at town 1 (GT) we see that town 3 is closest in terms of the formulation in the previous section, and we set $x_{13} = 1$. The closest town to 3 is 1, but town 1 has already been visited so we must visit the second closest, town 4. So we set $x_{34} = 1$. The closest to town 4 is town 5 and so $x_{45} = 1$. The closest town to 5 not visited is 8 so $x_{58} = 1$. Similarly, we ignore towns 4 and 5 when at 8 and set $x_{87} = 1$. From town 7 we go to town 9 and set $x_{79} = 1$. Next, x_{96} is set to 1. From town 6 we go to town 2 and set $x_{62} = 1$. When at town 2 the only unvisited town is 10. So we must set $x_{2,10} = x_{10,1} = 1$. The complete tour is $\langle 1, 3, 4, 5, 8, 7, 9, 6, 2, 10, 1 \rangle$. It has total distance 95 km and is shown in Fig. 4.9.

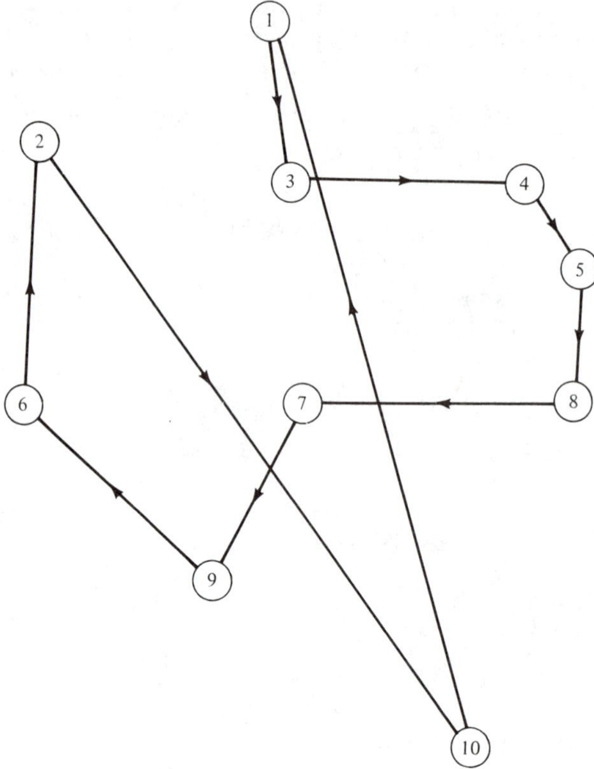

Figure 4.9. The Traveling-Salesman's Tour Derived by the Closest Neighbor Heuristic.

On looking at Fig. 4.9, it seems that town 10 has been responsible for the creation of a disastrous tour. Indeed the journeys in and out of 10 both cross other segments of the tour. It is well known in the theory of the traveling-salesman problem that the optimal tour does not possess such crossings. Thus we know that it must be possible to find an improved tour. In the next section we explain another heuristic which, at least for this problem, produces a better result.

4.2.3. The Closest Insertion Heuristic

This heuristic differs from the last in that it builds up a closed tour of towns. It begins at town 1. That is, the first tour is trivial—it comprises simply a tour in town 1, going nowhere else. At the next stage the town closest to 1 is identified as town 3. The first real tour, $\langle 1,3,1\rangle$, is created. Note that this is not a feasible solution as we are merely suggesting that the bus leaves 1, goes to 3, and then returns to 1. However, successive tours are built up in this way, each with one more town than the last. In the general step we ask what is the town not yet on the tour, which is the closest to a town on the tour? We then insert this town into the tour, taking as little extra distance as possible to do so. Let us illustrate this with the example. We now ask which town is closest to 1 or 3? The answer is that towns 4 and 7 are 7 km from town 3. Let us settle the tie by choosing 4 arbitrarily. We now insert town 4 into the tour to produce a new tour: $\langle 1,3,4,1 \rangle$.

Asking the same question we find that the closest town to the towns on the tour, namely, towns 1, 3, and 4, is town 5 which is 3 km from town 4. We now insert town 5 into the tour. There are three choices: We could insert it into any of the three journeys $1 \to 3$, $3 \to 4$, or $4 \to 1$. These insertions would produce the tours $\langle 1,5,3,4,1 \rangle$, $\langle 1,3,5,4,1 \rangle$, and $\langle 1,3,4,5,1 \rangle$, respectively. We choose the tour with the shortest total length. It is the second, $\langle 1,3,5,4,1 \rangle$, with distance 26 km. We do not have to add up the distance of all the journeys in each tour to find the shortest. It is far simpler to calculate the *extra* distance incurred by each insertion. Let us calculate the extra distance incurred by inserting 5 into the journey $1 \to 3$ to produce the sequence $1 \to 5 \to 3$; that is, the distance from town 1 to town 5 (12 km) plus the distance from town 5 to town 3 (9 km) less the distance from town 1 to town 3 (5 km) as the bus no longer has to travel directly from town 1 to town 3. Thus the extra distance is $12+9-5=16$ km. The extra distance for the insertion of town 5 into the journeys $3 \to 4$ and $4 \to 1$ is $d_{35}+d_{54}-d_{34}=9+3-7=5$ km and $d_{45}+d_{51}-d_{41}=3+12-9=6$ km, respectively. The correct choice will always be made if the new town is inserted so as to incur the least extra distance.

Continuing in this way, we find that town 8 is closest to towns on the tour, being 6 km away from town 5. The least-extra-distance insertion involves 8 being inserted on the $3 \to 5$ journey for an extra distance of $d_{38}+d_{85}-d_{35}=12+6-9=9$ km. As the previous tour, $\langle 1,3,5,4,1 \rangle$, had a length of 26 km this new tour, $\langle 1,3,8,5,4,1 \rangle$, has a length of $26+9=35$ km. Now town 7 is closest, being 7 km from town 3. It is best inserted on the $3 \to 8$ journey at an extra distance of $d_{37}+d_{78}-d_{38}=7+8-12$ km. This produces the tour $\langle 1,3,7,8,5,4,1 \rangle$ which has total distance 38 km. Next town 9 is inserted on the $7 \to 8$ journey at an extra distance of 9 km. The tour is now $\langle 1,3,7,9,8,5,4,1 \rangle$ and is of length 47 km. At the next stage both towns 2 and 6 are 8 km away from towns on the tour. We arbitrarily choose to insert 2. It is inserted on the $1 \to 3$ journey at an additional distance of 12 km. The tour is now $\langle 1,2,3,7,9,8,5,4,1 \rangle$ and is of length 59 km. Now town 6 is inserted on the $2 \to 3$ journey at an extra distance of 10 km. The tour is now $\langle 1,2,6,3,7,9,8,5,4,1 \rangle$ and is of length 69 km. Finally, town 10 is inserted on the $9 \to 8$ journey at an extra distance of 10 km. The final tour is now $\langle 1,2,6,3,7,9,10,8,5,4,1 \rangle$ and is of length 79 km. It is shown in Fig. 4.10. This is a substantial improvement over the tour produced in the last section.

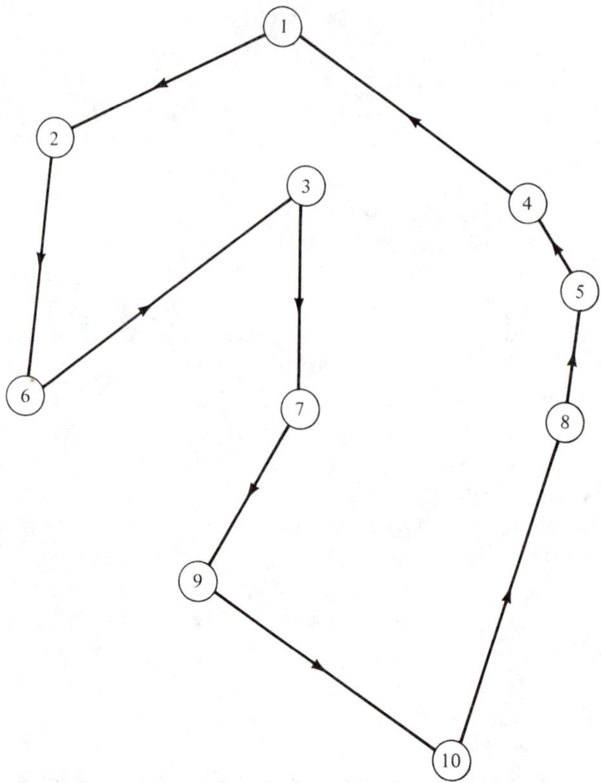

Figure 4.10. The Closest Insertion Heuristic Solution.

It can be seen from Fig. 4.10 that at least this tour does not possess any crossings, which would mean that it is, of necessity, suboptimal. We now discuss yet another heuristic which is somewhat more geometric in nature.

4.2.4. The Geometric Heuristic

Consider Fig. 4.11(a), where a boundary, called a *convex hull*, has been drawn around the towns. It is well known that the towns on the boundary will appear in any optimal tour in the order in which they appear on the convex hull. Thus we can consider the partial tour $\langle 1,2,6,9,10,8,5,4,1\rangle$ as being a good starting point to finding a complete tour of relatively low distance. All we have to do is to somehow include the "free" towns 3 and 7. The following heuristic inserts these cities with a view to minimizing the extra distance incurred, but in a different manner from the previous heuristic. A line is drawn from each of the free towns to each of the towns on the present tour of $\langle 1,2,6,9,10,8,5,4,1\rangle$. These lines are shown in Fig. 4.11(b). The next step is to calculate, for each free town, the angle between

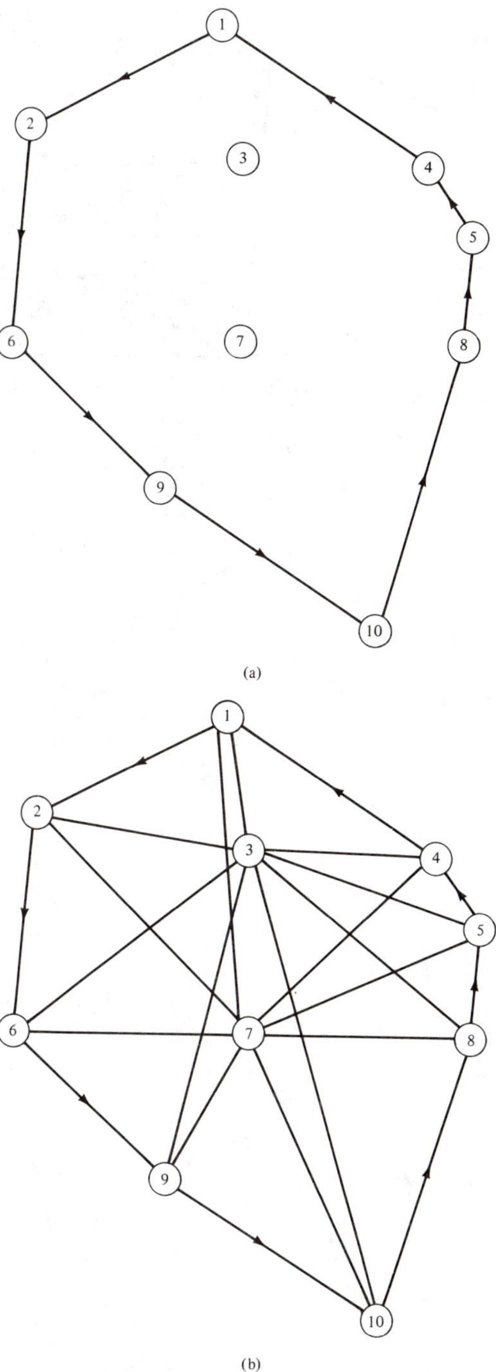

Figure 4.11. (a) The Convex Hull of the Towns; (b) The Angles Subtended by Towns 3 and 7 Which Are Not Yet Inserted.

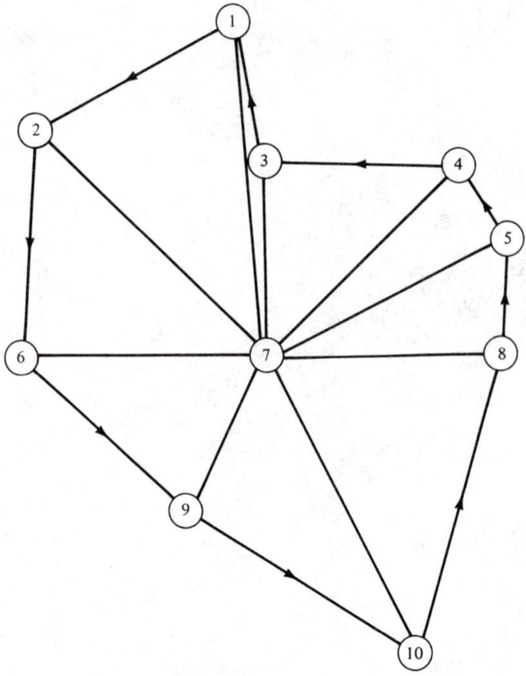

Figure 4.12. An Enlarged Tour ⟨1, 2, 6, 9, 10, 8, 5, 4, 3, 1⟩.

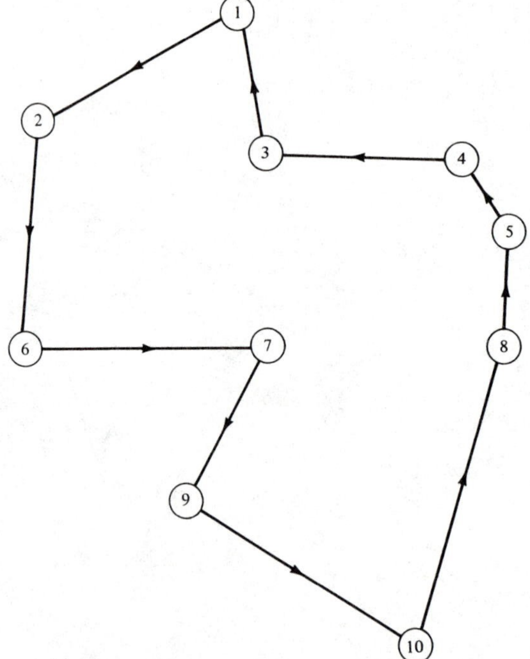

Figure 4.13. The Final Tour ⟨1, 2, 6, 7, 9, 10, 8, 5, 4, 3, 1⟩ of Length 67 km.

successive lines leaving it. The largest of all these angles is the one subtended by the line (4–1). It seems reasonable that the larger the angle the more sensible it is that the town at the apex of the angle be inserted on the journey between the two towns at the other end of the lines of the angle. Thus in this case as town 3 subtends the largest angle it is inserted first. Further, it is inserted in the intuitively obvious place between towns 1 and 4, with which it subtends the largest angle. This creates an enlarged tour $\langle 1,2,6,9,10,8,5,4,3,1\rangle$ which is shown in Fig. 4.12. The remaining free town, 7, is linked by straight lines to the towns on the tour and the largest angle is chosen. The largest angle for town 7 is subtended by line 6–9 and thus 7 is inserted on that journey. This produces the final tour $\langle 1,2,6,7,9,10,8,5,4,3,1\rangle$, which is of length 67 km and is shown in Fig. 4.13. This is a further improvement of the solution produced by the nearest-neighbor heuristic.

The geometric heuristic does not refer directly to the intertown distance table, but relies solely on the spatial arrangement of the towns.

4.2.5. Exercises

1. Solve the traveling-salesman problem with the following distance table using the nearest-neighbor heuristic:

	2	3	4	5	6	7	8	9	10
(GT) 1	6	8	4	3	9	12	14	20	1
2		11	5	6	8	3	2	1	5
3			14	16	18	29	10	9	8
4				5	12	14	40	30	25
5					17	16	15	18	22
6						12	14	30	3
7							4	9	12
8								8	7
9									2

2. Solve problem 1 above by using the closest insertion heuristic.

3. Consider the following 11 cities with coordinates: (1,4), (0,12), (9,3), (4,8), (6,2), (10,10), (8,4), (1,10), (9,6), (2,5), and (3,12). Calculate the straight-line distance between each pair. Use these distances and the geometric heuristic to find the shortest traveling-salesman tour through these cities.

4. Solve problem 3 above using the nearest-neighbor heuristic.

5. Solve problem 3 above using the closest insertion heuristic.

*6. Consider the optimal traveling-salesman tour $T=(v_1, v_2, \ldots, v_n)$, for n given points in the plane where the Euclidean metric is assumed. Over what region of the plane can any one of the given points in the set $\{v_1, v_2, \ldots, v_n\}$ be relocated if T is to remain optimal?

4.3. The Vehicle Scheduling Problem

4.3.1. Introduction

The SV Tourist Agency, introduced in Section 3.1, employs a printing company to produce its brochures. The company's press is located in Skier's Paradise (SP, town 7). Each month the company allocates its complete fleet of three small vans to the delivery of the brochures to the SV tourist agencies in the other nine towns. The brochures for SP are collected independently by the SP agency. The number of the bundles of brochures required by each of the towns per month is:

Town:	1	2	3	4	5	6	8	9	10
Required bundles:	10	15	18	17	3	5	9	4	6

The vans can each hold 40 bundles. What the company wants to do is to find a delivery schedule for each of the vans so that each van starts out at SP, visits a number of the other towns, delivers the bundles, and then returns to SP. Naturally the towns on any trip cannot have a total requirement of more than 40 bundles (van capacity). The objective is to find a schedule for each van so that each town is visited by exactly one van and the total distance traveled by all the vans is minimized. We calculate distances according to Table 3.1.

This problem is somewhat similar to the traveling-salesman problem, introduced in Section 4.2. Unfortunately, it is more complicated and there is even less hope of an efficient polynomial-time algorithm existing for it. Therefore we concentrate on the development of heuristic procedures, which are designed for large, realistic problems.

4.3.2. The Clark–Wright Heuristic

One of the most widely used vehicle scheduling heuristics was developed by Clarke and Wright. It begins by creating a separate tour for each town. That is, we create nine tours, each one leaving town 7, going directly to a different town, and then returning directly to town 7. These tours $(7, i, 7)$, $i = 1,2,3,4,5,6,8,9,10$, are shown in Fig. 4.14, which is based on the map of SV given in Fig. 3.1. Obviously, we have three times as many tours as we can handle and so these tours are combined to eventually produce three tours—one for each van.

Suppose we wish to combine the tours: $\langle 7,4,7\rangle$ and $\langle 7,5,7\rangle$. The truck would then leave town 7, visit towns 4 and 5, and then return to town 7. This would produce a new tour, either $\langle 7,4,5,7\rangle$ or $\langle 7,5,4,7\rangle$. We use the

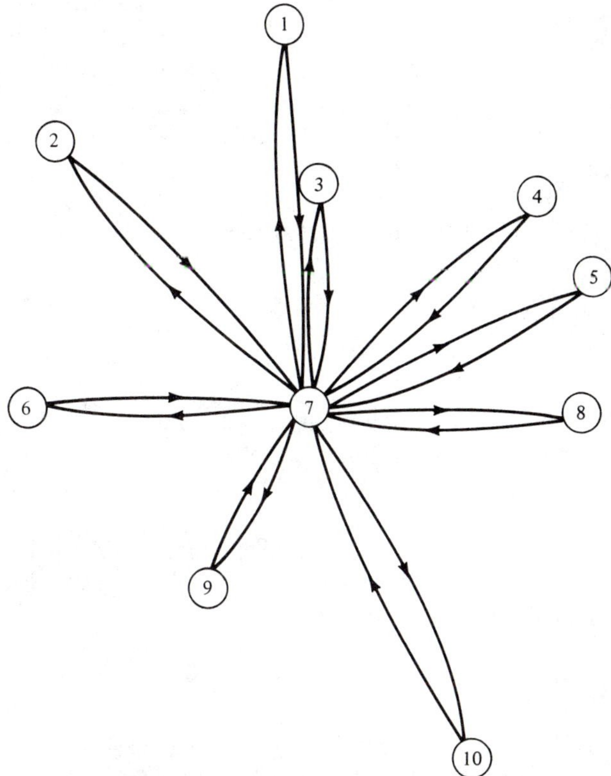

Figure 4.14. Initial Tours for the Vehicle Scheduling Problem According to the Clarke–Wright Heuristic.

intertown distance table given in Table 3.1 to discover whether either of these modifications are advantageous. Because the distance d_{ij} from town i to town j equals the distance d_{ji} in the reverse direction, the total distance of these tours is the same. Therefore we adopt the first tour: $\langle 7,4,5,7\rangle$. How much does this combination of tours save over individual tours? The individual tours $\langle 7,4,7\rangle$ and $\langle 7,5,7\rangle$ have total length

$$d_{74}+d_{47}+d_{75}+d_{57}=2(d_{74}+d_{75}).$$

The combined tour $\langle 7,4,5,7\rangle$ has length

$$d_{74}+d_{45}+d_{75}.$$

Thus the combining of the tours has saved

$$2(d_{74}+d_{75})-(d_{74}+d_{45}+d_{75})=d_{74}+d_{75}-d_{45}. \tag{4.1}$$

If

$$d_{74}+d_{75}-d_{45}>0,$$

it is a good idea to combine the tours in this way as it saves distance. The only other matter to check is that the total requirement of towns 4 and 5 on the combined tour does not exceed van capacity. First,

$$d_{74}+d_{75}-d_{45}=10+10-3=17>0,$$

so distance would be saved by this combination. Second, the demands of towns 4 and 5 are 17 and 3, respectively, so the total demand of 20 is well below van capacity of 40. So in light of this it seems a good idea to combine the two trips as its saves distance and is feasible.

However, is this the best combination to make? Are there other combinations which are also feasible and which save more distance? Which is the feasible combination which saves the most distance? It is clear from Eq. (4.1) that if tours (7, i,7) and (7, j,7) to towns i and j are combined then the savings in distance is

$$s_{ij} = 2(d_{7i} + d_{7j}) - (d_{7i} + d_{7j} + d_{ij}) = d_{7i} + d_{7j} - d_{ij}.$$

We denote this quantity by s_{ij}. Note that s_{ij} is not necessarily positive for all $i-j$ pairs.

Rather than make the combination of tours to towns 4 and 5 we return to the original solution in Fig. 4.14 and calculate s_{ij} for all $i-j$ pairs. These are shown in Table 4.4. The values should be calculated independently by the reader.

It turns out that the previously discussed combination yields the highest saving so this combination is made. The total distance traveled for the original schedule, shown in Fig. 4.14, is

$$2\sum_{\substack{i=1\\ i\neq 7}}^{10} d_{7i} = 2(12+11+7+10+10+9+8+6+11) = 168 \text{ km}.$$

With the combined tour $\langle 7,4,5,7\rangle$ saving 17 km, the new total distance is $168-17=151$ km. We now make further combinations, in order of largest feasible saving first.

The next largest saving involves combining the trips to towns 1 and 2 for a saving of $s_{12}=15$ km. As the requirements of towns 1 and 2 sum to $10+15=25$ bundles, this combination is feasible and is made. We have now created the tour $\langle 7,1,2,7\rangle$. The tour distance traveled is now $151-s_{12}$ $=151-15=136$ km.

The next largest saving involves combining the trips to towns 1 and 3. However, the total requirement of the towns 1, 2, and 3 is $10+15+18=43$ bundles, which exceeds van capacity. So this saving cannot be realized and

Table 4.4

		2	3	4	5	6	8	9	10
(GT)	1	15	14	13	10	7	4	1	1
	2		9	6	4	12	1	3	0
	3			10	8	5	3	1	1
	4				17	2	11	1	3
	5					1	12	1	6
	6						3	7	4
	8							3	8
	9								7

we turn to the next largest saving, which involves $s_{25}=13$. This implies combining the tour of which 2 is a part, $\langle 7,1,2,7\rangle$, and the tour of which 5 is a part, $\langle 7,4,5,7\rangle$. Unfortunately, the total demand of towns 1, 2, 4, and 5 is $10+15+17+3=45$, which exceeds van capacity, so this combination cannot be made. There are two possible savings which are the next largest: $s_{58}=s_{26}=12$. Dealing arbitrarily with s_{58} first, we see that this suggests combining the tour to 5 with the tour to 8. There is a complication in that 5 is already on a combined tour: $\langle 7,4,5,7\rangle$. If we extend this tour to include 8 in the obvious way we produce the tour $\langle 7,4,5,8,7\rangle$. Is the saving really 12 as advertised? We originally had the distance $d_{74}+d_{45}+d_{57}$ (for the first tour) and $2d_{78}$ (for the second tour) making a total of $d_{74}+d_{45}+d_{57}+2d_{78}$. After the combination, the total distance is $d_{74}+d_{45}+d_{58}+d_{87}$.

The difference between the two is

$$d_{74}+d_{45}+d_{57}+2d_{78}-(d_{74}+d_{45}+d_{58}+d_{87})=d_{57}+d_{78}-d_{58}=12,$$

which equals s_{58}. As the requirements of the towns on the new tour is $17+3+9=29$, which is less than van capacity, this combination is made.

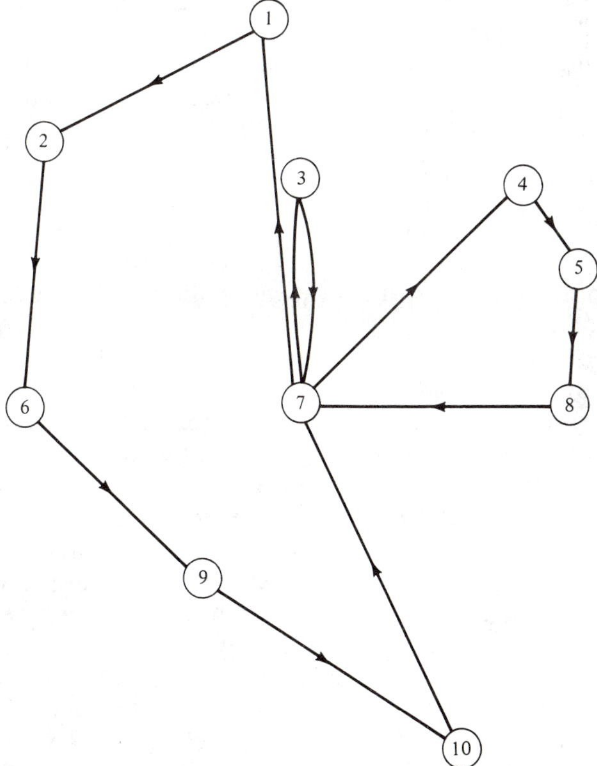

Figure 4.15. The Final Delivery Schedule for the Vans Derived by the Clarke–Wright Heuristic.

This produces a current total distance traveled of $136 - s_{58} = 136 - 12 = 124$ km. The other saving of 12, s_{26}, can also be made and it produces the tour $\langle 7,1,2,6,7 \rangle$ with the total requirement of $10+15+5=30$. This produces a current total distance traveled of $124 - s_{26} = 124 - 12 = 112$ km.

The next largest saving $s_{48} = 11$ can be overlooked as 4 and 8 are already on the same tour. There are two savings of value 10; s_{15} and s_{34}. Both are infeasible.

The next few savings s_{23}, s_{35}, and $s_{8,10}$ are all infeasible and we come to $s_{16} = s_{69} = s_{9,10} = 7$. The first is inappropriate as 1 and 6 are on the same tour. However, the other two are feasible and can be made to produce first the tour $\langle 7,1,2,6,9,7 \rangle$ and then the tour $\langle 7,1,2,6,9,10,7 \rangle$. No further savings can be made and we are left with the following final schedule:

Tour	Distance (in km)	Total requirement (brochure bundles)
$\langle 7,3,7 \rangle$	14	18
$\langle 7,4,5,8,7 \rangle$	27	29
$\langle 7,1,2,6,9,10,7 \rangle$	57	40

This schedule, which has a total distance of 98 km, is shown in Fig. 4.15.

The configuration in Fig. 4.15 appears somewhat unsatisfactory. And yet the problem is tightly constrained by van capacity, making more efficient town groupings not easy to find. We turn now to a more intuitive, geometric heuristic which, at least for this problem, produces equivalent results.

4.3.3. The Sweep Heuristic

The sweep heuristic, to be introduced now, differs substantially from the last-mentioned heuristic. It does not build up a set of tours by combination, but partitions the towns into subsets with each subset making up one tour. Once the towns on a tour are all known, the actual tour itself can be found by applying a traveling-salesman heuristic (see Section 4.2). The method begins by passing an arc through town 7 directly northwards, as shown in Fig. 4.16. Next this arc is swung in a clockwise direction with town 7 as the pivot. A record is kept of the order of the towns through which it passes. This order is $3,4,5,8,10,9,6,2,1$. Towns are grouped according to this order on the basis of requirement. That is, the first van will have on its schedule towns in this order until it cannot service a further town because of its load capacity. It then returns to town 7. The next van carries on where the last van left off and so on. Thus the first van visits towns 3, 4, and 5, at which point it has no bundles left to deliver to 8, the next town on the list. So it returns to town 7. The next van visits 8, 10, 9, 6, and 2. The last truck visits town 1 only. So the towns are clustered into the sets: $\{3,4,5\}$, $\{8,10,9,6,2\}$, and $\{1\}$, as shown in Fig. 4.16. The problem of finding the actual sequence of towns to be visited by a van in each group is a traveling-salesman problem as van capacity is no longer an issue. Applying a traveling-salesman heuristic, we discover that the least-distance tours for the clusters turn out to be to visit the towns in the order in which they were passed over by the ray. This is not always the case. The tours are shown below:

Tour	Distance (km)	Total requirement (brochure bundles)
⟨7,3,4,5,7⟩	27	38
⟨7,8,10,9,6,2,7⟩	56	39
⟨7,1,7⟩	24	10

This first schedule has a distance of 107 km, which is worse than that produced in the last section.

Next, a ray is passed in an easterly direction through town 7 just above town 8. The above process is repeated, producing the following clusters:

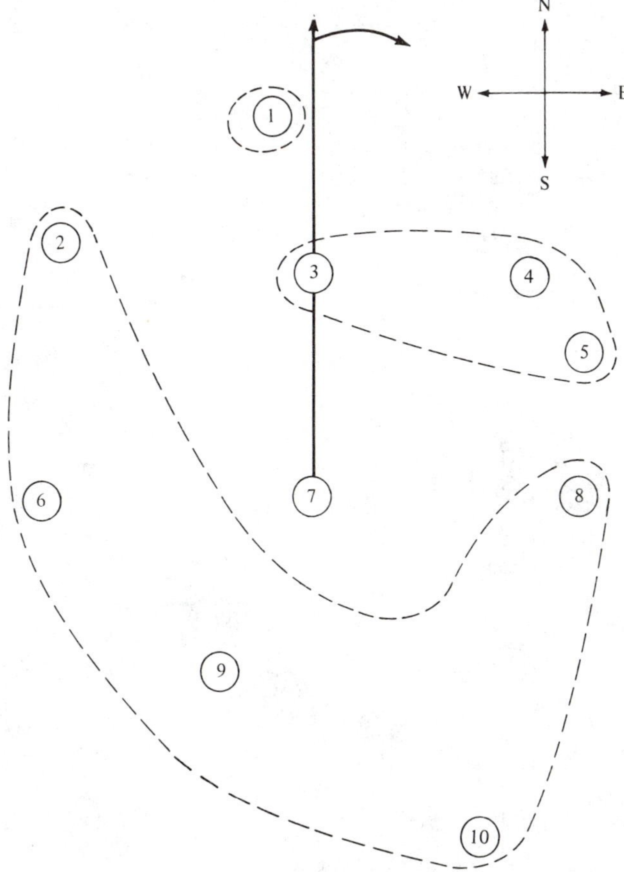

Figure 4.16. The Initial Clustering of the Towns as Derived by the Sweep-Heuristic Method.

(8, 10, 9, 6, 2), (1, 3), and (4, 5). This time the tours are:

Tours	Distance (km)	Total requirement (brochure bundles)
⟨7, 8, 10, 9, 6, 2, 7⟩	56	39
⟨7, 1, 3, 7⟩	24	28
⟨7, 4, 5, 7⟩	23	20

This schedule has a total distance of 103 km which is once again an improvement.

The ray is now passed in a westerly direction through town 7 and the process is repeated. This produces the following tours:

Tours	Distance (km)	Total requirement (brochure bundles)
⟨7, 6, 2, 1, 7⟩	37	30
⟨7, 3, 4, 5, 7⟩	27	38
⟨7, 8, 10, 9, 7⟩	35	19

This schedule has total distance 99 km. Finally, each of the four rays is moved in an anticlockwise direction to produce four new schedules. The schedules in order of the northerly, easterly, southerly, and westerly rays are:

	Tours	Distance (km)	Total requirement (brochure bundles)
Northerly	⟨7, 1, 2, 6, 9, 10, 7⟩	57	40
	⟨7, 8, 5, 4, 7⟩	27	29
	⟨7, 3, 7⟩	14	18
		98	
Easterly	⟨7, 5, 4, 3, 7⟩	27	38
	⟨7, 1, 2, 6, 9, 10, 7⟩	57	40
	⟨7, 8, 7⟩	16	9
		100	
Southerly	⟨7, 10, 8, 5, 4, 7⟩	41	35
	⟨7, 3, 1, 7⟩	24	28
	⟨7, 2, 6, 9, 7⟩	33	24
		98	
Westerly	⟨7, 9, 10, 8, 5, 4, 7⟩	46	39
	⟨7, 3, 1, 7⟩	24	28
	⟨7, 2, 6, 7⟩	28	20
		98	

The northerly solution is the same as the best solution found by using a clockwise sweep. There are two other solutions (the southerly and westerly solutions) which also have this value. This sweep heuristic has an advantage over the first heuristic in that it produces many solutions. This may be of use when there are further constraints on the tours.

4.3.4. Exercises

1. Consider problem 1 of Section 4.2.5 as a vehicle scheduling problem where town i has a demand for i units. Van capacity is 20 and no more than three vans are to be used. Solve this problem by the Clarke–Wright heuristic.
2. Solve problem 1 above by the sweep heuristic.
3. Consider problem 3 of Section 4.2.5 as a vehicle scheduling problem where point i has a demand for i units. Van capacity is 20 and no more than four vans are to be used. Solve this problem by the Clarke–Wright heuristic.
4. Solve problem 3 above by the sweep heuristic.

*5. Show how to modify the sweep heuristic so that it can solve problems where each van has different capacities.

4.4. Car Pooling

4.4.1. Introduction

The SV Tourist Agency, introduced in Section 3.1, has monthly meetings which are held in Geyser Town. One agent from each of the other towns travels by car to the meetings. In the past each agent has taken his or her own car. The soaring price of gasoline has recently prompted the GT Agency to issue a memo suggesting that car pooling be used. That is, certain agents will pick up others on their way to GT in order to save running costs. Some agents may be inconvenienced in the sense that they have to travel slightly further. This is compensated by the fact that most people no longer have to drive and also much gasoline is saved. It is assumed that a number of drivers can meet at another agent's home and leave all but one car there before picking up the agent and driving on in one car. Cars cannot be left at points other than agent's homes. It is further assumed that all cars have the capacity to accommodate a driver and four other agents.

4.4.2. The Problem in Graph Theoretic Terms

We now describe the problem of finding the least-distance car-pool policy and a solution procedure for it in terms of graph theory (see Chapter 5, Section 5.2 for an explanation of terminology). Each of the towns is represented by the vertex of a graph. Each positive, finite $i-j$ entry in the upper triangle of a distance table is the distance of an actual road linking towns i and j. It is associated with an edge joining vertices i and j in the underlying graph. We denote by G the graph of these vertices and edges. The problem is to identify a collection $C=(T_1, T_2, \ldots, T_m)$ of trees, where each tree T_i, $i=1,2,\ldots,m$ is a subgraph of G and:

(1) Each vertex, other than GT, is a member of exactly one tree.
(2) Vertex 1 is a member of every tree.
(3) Each tree has five or less edges.
(4) The sum of the distances of all the edges in all the trees is a minimum.

The sum mentioned in (4) above is called the *cost* of the collection C. Each tree T_i is called a *trip*. Restriction (1) represents the fact that a driver sets out from each of the homes located at the pendant vertices of the tree. Each of these drivers drives along the roads represented by the path incident with the pendant vertex. If a vertex p of degree three or more is reached before point 1, the driver waits there until all other cars which must pass through p arrive. Then all the drivers and the agent who lives at p proceed towards 1 in one of the cars, leaving the other car or cars at p. Eventually, all the agents arrive at 1, which explains restriction (2). Because each car cannot carry more than five agents each trip cannot have more than five edges, which accounts for restriction (3). The numerical example in the next section should make these ideas more clear.

It is possible to formulate the problem as an integer programing problem, the solution of which guarantees minimum distance. Even with a modestly sized problem of about 20 agents, the number of constraints is enormous. Thus this approach does not seem practical on present-day computers. Moreover, any method which will be widely used must be amenable to quick and simple hand calculation and not require advanced mathematics or a computer. Therefore we develop three *heuristic* procedures (see Chapter 2 for an explanation of this term) in the next sections.

4.4.3. The Nearest-Point Procedure

This method is designed to be carried out by sticking pins in maps and by comparing distances by using a piece of string on a map or by simple estimation by eye. The approach means that it is not necessary to construct a distance table. However, we use a distance table in the numerical example in order to explain the method easily. The heuristic is constructive in the sense that it iteratively builds up the final solution, edge by edge. We consider each of the towns $2,3,\ldots,10$ in turn, in the order of distance from GT, farthest away from GT first. In the general step we accept as part of the solution the edge out of vertex i which is of least distance such that: (i) it does not create a cycle with edges already chosen, (ii) a trip cannot have more than five edges, and (3) each trip must contain vertex 1. These rules mean that sometimes the only edge out of vertex i that can be accepted is $\{i,1\}$, the edge joining i to GT. For instance, if a trip already has four edges then the only legitimate edge to accept is $\{i,1\}$. The heuristic is terminated when every vertex except 1 has been considered. It will now be illustrated by solving the Scenic Valley problem whose distance table is Table 4.5. It is an extension of Table 3.1.

Table 4.5

		Towns									
		1	2	3	4	5	6	7	8	9	10
	2	8	0	9	15	17	8	11	18	14	22
	3	5	9	0	7	9	11	7	12	12	17
	4	9	15	7	0	3	17	10	7	15	18
Towns	5	12	17	9	3	0	18	10	6	15	15
	6	14	8	11	17	18	0	9	14	8	16
	7	12	11	7	10	10	9	0	8	6	11
	8	16	18	12	7	6	14	8	0	11	11
	9	17	14	12	15	15	8	6	11	0	10
	10	22	22	17	18	15	16	11	11	10	0

The first row is missing because we are concerned only with the travel of the agents to GT (town 1). The policy for the return journeys is deduced simply by reversing the direction of travel in the original solution.

The heuristic begins by listing the towns in order of decreasing distance from GT. Scanning the first column of Table 4.5, we see that this order is 10, 9, 8, 6, 7, 5, 4, 2, 3. Towns 5 and 7 are both 12 km from 1 and their order in the above list is settled arbitrarily. Here we settle ties on the basis of the town with largest number being selected first.

We consider town 10 first. Scanning the bottom row of Table 4.5 we see that it is closest to town 9. Thus edge $\{10,9\}$ is accepted as part of the solution. The solution being built up is shown in Fig. 4.17. The reader is urged to make a copy of Fig. 4.17 and build up the solution independently. Next town 9 is considered, not because of the edge $\{10,9\}$, but because it is the next on the list. Scanning row 9 we see that it is closest to town 7 and thus edge $\{9,7\}$ is selected. Town 8 is closest to town 5 and edge $\{8,5\}$ is selected. Town 6 is as close to town 9 as to any other town and edge $\{6,9\}$ is selected. Town 7 is closest to town 9, but edge $\{7,9\}$ is not selected as it would create the cycle $\langle 7,9,7\rangle$ as edge $\{7,9\}$ has already been chosen. Thus the next closest town to 7 is chosen, town 3, and edge $\{7,3\}$ is selected. At this point the reader should have in his or her diagram five edges. Ignoring edge $\{8,5\}$, we have a tree (representing a trip) connecting vertices 6, 9, 10, 7, and 3. As there are five vertices representing five towns with five agents in the trip, edge $\{3,1\}$ must be accepted automatically. This is because we have assumed that car capacity is five agents. Noting the directions on the edges in Fig. 4.17 we see that drivers 10 and 6 both drive to town 9. One car is left there, and agent 9 is picked up in the other. The three agents then drive in

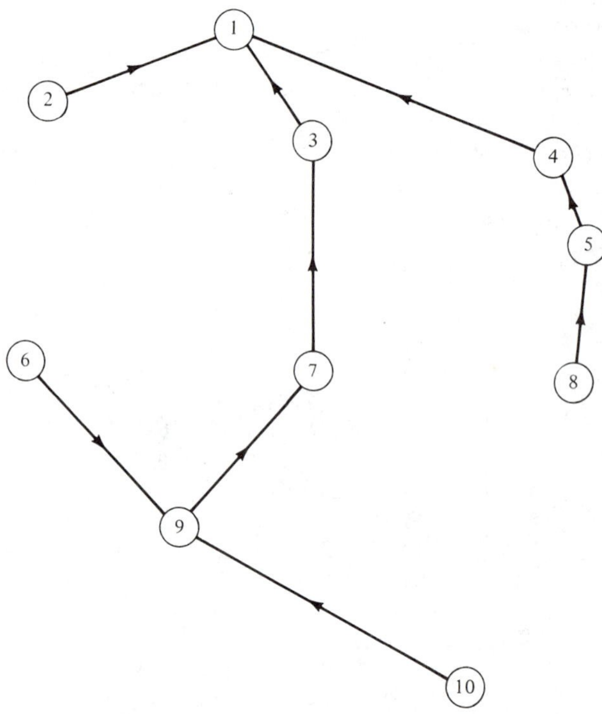

Figure 4.17. The Nearest-Neighbor-Heuristic Solution.

one car to town 3 picking up agent 3. At this point the car is full so the only possibility is to drive directly to town 1.

Continuing with the heuristic, town 5 is the next on the list and it is closest to town 4. Edge $\{5,4\}$ is accepted. Town 4 is closest to town 5 but edge $\{4,5\}$ is rejected as it would create a cycle. The next closest towns are 3 and 8. But acceptance would create either an overloaded car at town 3 (a tree with more than five edges) or the cycle $\langle 4,5,8,5 \rangle$. So edge $\{4,1\}$ is accepted. The complete solution is illustrated in Fig. 4.17. It comprises three trips. The first trip, involving agents 6, 9, 10, 7, and 3, has already been described. In the second trip, agent 8 picks up agent 5 and then agent 4 before driving to GT. In the third trip, agent 2 drives directly to GT.

The total distance traveled is the sum of the distances associated with the edges in Fig. 4.17, namely, 62 km. The total distance traveled if all the agents travel independently is the sum of the first column of Table 4.5, namely, 115 km. So nearly half the total distance traveled has been saved.

4.4.4. The Triangular Heuristic

Like the nearest-point heuristic, the triangular heuristic is also constructive in the sense that it also builds up a solution edge by edge. When it is implemented, the following steps are carried out (where d_{ij} denotes the distance between towns i and j):

(1) Let p be the vertex not yet considered which is the farthest away from 1.
(2) List all edges $\{p, j\}$ with the properties:
    (a) $d_{j1} \leq d_{p1}$.
    (b) $d_{pj} < d_{p1}$.
(3) Accept as part of the solution, if possible, edge $\{p, j\}$ which is of the shortest distance among all those found in step (2) and which does not create a cycle or a trip with more than five edges. If there is no such edge, accept edge $\{p,1\}$ and go back to step (1). If there is such a suitable edge set p to become j and go back to step (2).
(4) Repeat steps (1), (2), and (3) until all vertices belong to one trip each.

The heuristic will now be illustrated by using it to solve the SV problem. The solution built up is illustrated in Fig. 4.18. Vertex 10 maximizes d_{i1} so we begin here. Edge $\{10,9\}$ is the shortest out of 10 and is accepted. We then implement steps (2) and (3) above and accept $\{9,7\}$ by taking the least-distance edge from 9, which joins to a vertex no further from 1 (namely, vertex 7). Repeating this process, we would accept $\{7,9\}$ again but it creates a cycle. So we accept $\{7,3\}$ as 3 is closer to 1 than 7. We accept $\{3,1\}$ as 1 is closest to 3. We next consider 8, which is the farthest vertex from 1 not yet considered, and accept edge $\{8,5\}$ as 5 is closer to 1 than 8

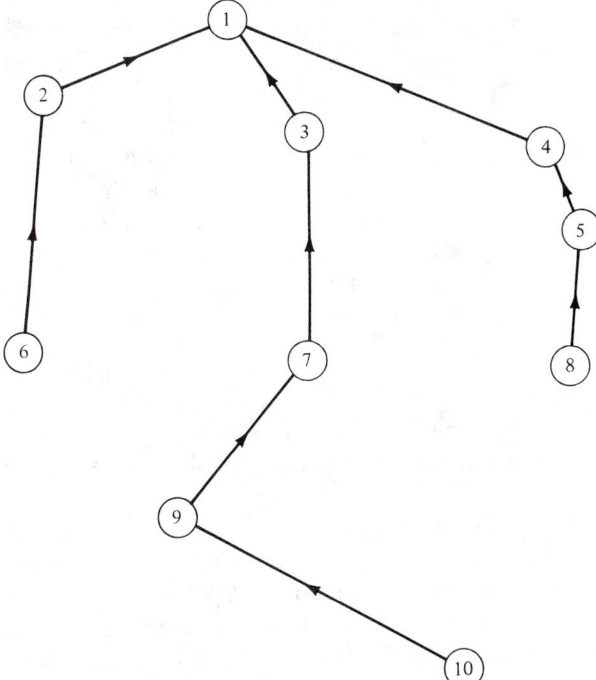

Figure 4.18. The Triangulation-Heuristic Solution.

is. Similarly, we accept $\{5,4\}$. At the next iteration we would like to accept $\{4,5\}$ or $\{4,8\}$, but this would create a cycle. We would then like to accept $\{4,3\}$, but this would create a trip with more than five edges. We end up accepting $\{4,1\}$. The next farthest vertex is 6. It is equidistant from 9 and 2. Vertex 2 is closer to 1 than 6 is, vertex 9 is not closer to 1 than 6 is. Therefore $\{6,2\}$ must be chosen as $\{6,9\}$ cannot be considered. Finally, $\{2,1\}$ is chosen. This solution has total distance 61 km, 1 km less than the nearest-neighbor solution.

4.4.5. The Tree-Collapsing Heuristic

Unlike the heuristics of the previous two sections, this heuristic is an improvement procedure. That is, it takes an initial solution to the problem (not necessarily one that satisfies car capacity) and systematically improves it. The initial solution is provided by the triangular heuristic with car capacity temporarily ignored. This solution is then modified in the following manner.

We associate with each town (other than town 1) the number of agents who would leave that town if the solution was adopted. If no town receives a number greater than five, then the solution is feasible and we adopt it. Otherwise, we modify the solution as follows:

(1) If some town i has a number k and the edge out of vertex i in the solution does not join it to town 1, we delete the edge out of i and accept edge $\{i,1\}$ instead.
(2) Suppose some town j has a number greater than k, but none of the edges into j begin at towns with numbers greater than or equal to k. We choose from these towns a collection U, such that the sum of their numbers is as great as possible, but no more than $k-1$. Retain the edges linking these towns to j. Let the towns not involved in the sum form the set W. Then we form new trips by deleting the edges from towns in W to j. If the edge out of j is (j,s) where $s \neq 1$, delete (j,s) and join each town in W to town 1.

Performing all changes under rules (1) and (2) above, we create a new collection of trips. We number the towns anew, and, if necessary, apply rules (1) and (2) repeatedly. We continue until all trips have five or less edges.

The SV problem is now solved using this heuristic. The initial solution is shown in Fig. 4.19. The number of agents leaving each town is shown. There are just two trips. The simple trip involving agent 2 alone is feasible and is not touched. It is going to be crowded in the car leaving the house of agent 3 and we must do something about this. There is no town with a number 5, so we cannot invoke rule (1). We come to rule (2). Town 3 has a

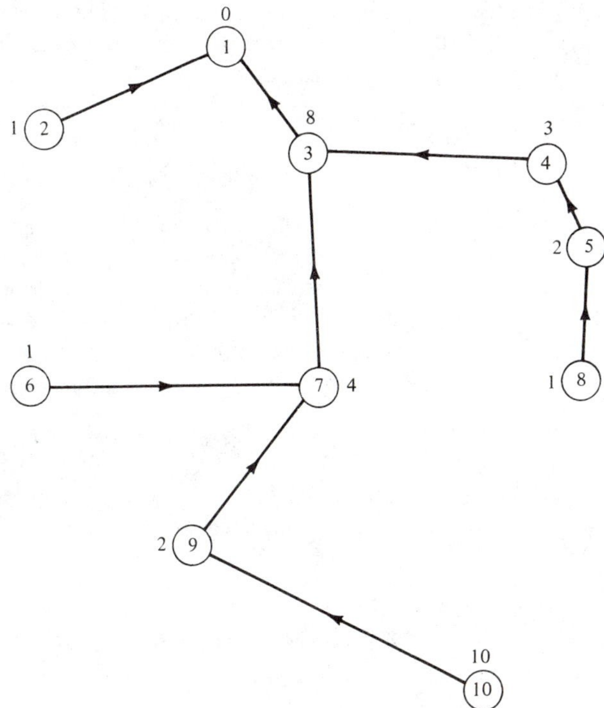

Figure 4.19. The Initial Solution for the Tree-Collapsing Heuristic.

number greater than 5, namely, 8. So we set $j = 3$. The towns with edges going into 3 are 4 and 7 and neither has a number greater than 5. We must choose a set U from 4 and 7 so that the sum of their numbers is as great as possible, but does not exceed 5. Towns 4 and 7 have numbers 3 and 4, respectively. So, of course, we cannot have both 4 and 7 in U because the numbers sum to $3+4=7$, which is greater than 5. So the closest we can get to 5 from below is to set $U = \{7\}$. The edge $\{7,3\}$ is then retained. The town not involved in the sum is 4 and thus we set $W = \{4\}$. We delete $\{4,3\}$ and select $\{4,1\}$ instead. This creates the final solution, which is identical to that obtained by the nearest-point heuristic.

4.4.6. A Comparison of the Three Heuristics

The heuristics of the three previous sections all gave solution values for the SV problem which are very similar. For large problems the first two heuristics usually produce solutions about 3% above the optimum and the tree-collapsing procedure about 6%. Problems with 20 to 30 people to be picked up can usually be solved by hand calculation in under 5 min, working on a scale map with pins and string. As the triangular heuristic is usually quicker to implement than the nearest-point procedure, it is probably the most useful in any practical exercise.

4.4.7. Improvements to the Solution

Quite often the solutions obtained by the previous heuristics can be seen at once to be suboptimal. We would naturally like to incorporate any improvements we can easily see and it may be worthwhile to conduct a brief search. It is not easy to specify all possible kinds of improvement which might be found but the following rules, especially the first, are easily seen and used. They all assume that the change suggested does not violate car capacity:

(1) If there is an edge from i to j and $d_{ir} < d_{ij}$, delete $\{i, j\}$ and insert $\{i, r\}$.

(2) If a trip begins at town r with edge $\{r, s\}$ and there is an edge $\{i, j\}$ (not necessarily part of the same trip) such that

$$d_{ir} + d_{rj} < d_{ij} + d_{rs},$$

delete $\{i, j\}$ and $\{r, s\}$ and insert $\{i, r\}$ and $\{r, j\}$.

(3) If $\{i, r\}$ and $\{r, j\}$ belong to the same trip and s is another town such that

$$d_{ij} + d_{rs} < d_{ir} + d_{rj},$$

delete $\{i, r\}$ and $\{r, j\}$ and insert $\{i, j\}$ and $\{i, s\}$.

4.4.8. Exercises

1. Solve the following car-pool problem by all three heuristics where car capacity is 7 and the meeting point is denoted by 0.

	0	1	2	3	4	5	6	7	8	9	10	11	12	13	14	15
1	1	0	5	2	6	4	2	2	3	3	3	4	2	4	3	4
2	6	5	0	3	1	1	5	7	5	8	9	9	7	8	8	9
3	3	2	3	0	4	2	4	4	5	5	6	6	4	5	5	6
4	7	6	1	4	0	2	4	8	9	9	10	10	7	8	8	9
5	5	4	1	2	2	0	6	6	7	7	8	8	6	7	7	8
6	2	2	5	4	4	6	0	4	5	5	6	6	3	4	4	5
7	1	2	7	4	8	6	4	0	1	1	2	2	2	3	3	6
8	2	3	5	5	9	7	5	1	0	2	1	3	3	4	4	5
9	2	3	8	5	9	7	5	1	2	0	3	1	3	4	4	5
10	3	3	9	6	10	8	6	2	1	3	0	4	4	5	5	6
11	3	4	9	6	10	8	6	2	3	1	4	0	4	5	5	6
12	1	2	7	4	7	6	3	2	3	3	4	4	0	1	1	2
13	2	3	8	5	8	7	4	3	4	4	5	5	1	0	2	1
14	2	3	8	5	8	7	4	3	4	4	5	5	1	2	0	3
15	3	4	9	6	9	8	5	4	5	5	6	6	2	1	3	0

2. Improve the three solutions found to problem 1 above by using the techniques mentioned in Section 4.4.7.

*3. Explain how the heuristics can be modified to account for the possibility that more than one agent travels from each town.

*4. Explain how the heuristics can be modified to account for the situation of finding a good solution for the homeward journeys when $d_{ij} \neq d_{ji}$ for some towns i and j.

4.5. Evolutionary Tree Construction

4.5.1. Introduction

In this section we turn to a problem in biology. The theory of evolution continues to be a focus for nearly all biological thought. According to current theories of evolution, existing biological species have been linked in the past by common ancestors. Following the Darwinian school, many scientists have represented postulated ancestral relationships by trees called *phylogenies*. The ancestors of certain groups of species, such as vertebrates, have left a rich fossil record of their existence, which can be used to make comparisons with similar existing species. This has lead to a fair degree of agreement on the structure of phylogenies of these groups. Unfortunately, for most groups the record is inadequate and in some cases unknown. In these cases there is often considerable disagreement over the nature of the phylogenies which describe their histories.

Each generation of a species passes on to the next generation a very detailed account of itself, written in the chemical code of DNA from which genes are constructed. Therefore a comparative analysis of the equivalent genes, or the proteins for which they code, from different species should reveal evidence of the ancestral relationship of the species.

Over the past two decades attempts have been made to construct phylogenies from protein sequence data. These methods typically construct a phylogeny for a particular set of species given a unique protein sequence for each of the member species. It is assumed that all sequences are derived from the same protein—typically the respiratory protein cytochrome *c*. The symbols in each sequence normally represent amino acids. Each amino acid can be represented by an ordered triple of nucleotides. Since there are four nucleotides, A, C, G, and U, there are a priori 64 possible triples. However, only 20 of these have been found to occur in nature. Thus each sequence comprises a string of symbols from an alphabet of size 20.

Several workers in the field have found it advantageous to convert amino acid sequences into nucleotide sequences for the construction of phylogenies. This is because differences between different pairs of amino acids can vary, whereas differences between different pairs of nucleotides can conveni-

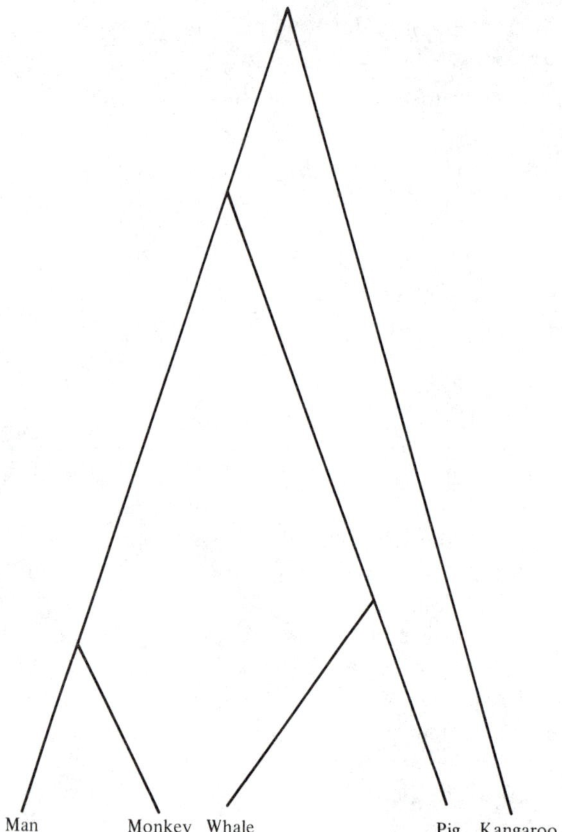

Figure 4.20. A Directed Phylogeny.

ently all be assigned unit weight. Henceforth we shall assume that the data have undergone such a transformation.

The basic objective of this approach is to construct a phylogeny in which each given species and its sequence is represented. It is usual to first construct an *unrooted* phylogeny, an example of which is given in Fig. 4.20. An unrooted phylogeny does not have the point representing the common ancestor of all the given species distinguished. A common ancestor is then specified by *directing* the phylogeny, that is, by giving each edge in the tree an orientation directed away from the common ancestral point, as illustrated in Fig. 4.21.

In this chapter we shall be concerned only with the construction of undirected phylogenies. The endpoints of each edge in the phylogeny represent nucleotide sequences, which can be examined at each site for differences. The number of sites at which differences occur is associated with the edge. A commonly used optimality criterion (which we also use) is to minimize the sum of these numbers taken over all edges. A phylogeny selected under this criterion is said to be of *maximum parsimony*. We now make these ideas more precise.

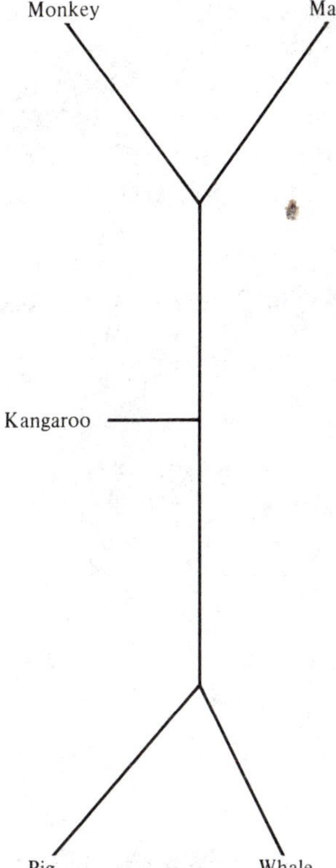

Figure 4.21. An Undirected Phylogeny.

4.5.2. A Mathematical Model

We now introduce a numerical example in order to explain the concepts of the previous section. Consider a given set of species s_1, s_2, s_3, s_4, and s_5 with nucleotide sequences given in Table 4.6.

In most realistic biological studies, the number of species and the sequence length are at least 30 and 100, respectively. However, we consider a much reduced example here for expository reasons.

Table 4.6. The Nucleotide Sequences for Five Species

		Sites							
		1	2	3	4	5	6	7	8
Species	s_1	A	G	U	G	U	U	A	A
	s_2	C	A	A	G	U	U	A	A
	s_3	C	G	U	C	C	G	A	A
	s_4	G	A	U	G	U	U	C	A
	s_5	C	G	U	G	U	U	C	U

What we want to do is to connect all the species by a tree with the minimum possible total number of changes associated with its edges. The general problem is similar to the minimal-spanning-tree problem (MSTP) of Chapter 3. If we treat our problem as an MSTP we can form a table of the number of differences between each pair of species. This is given as Table 4.7. As an example, the $s_1 - s_2$ entry is 3 because the sequences (for sites 1 to 8)

Table 4.7. The Number Differences between Species

		Species			
		s_2	s_3	s_4	s_5
	s_1	3	4	2	3
	s_2		5	3	4
Species	s_3			5	5
	s_4				3

$$\begin{array}{cccccccc} 1 & 2 & 3 & 4 & 5 & 6 & 7 & 8 \\ s_1 = A & G & U & G & U & U & A & A \end{array}$$

and

$$\begin{array}{cccccccc} s_2 = C & A & A & G & U & U & A & A \end{array}$$

differ at three sites, namely, sites 1, 2, and 3. The MSTP is the problem of connecting the species by a tree with the minimum total number of differences. If we apply one of the MSTP algorithms of Section 3.1 to the data of Table 4.7 we can produce the minimal spanning tree shown in Fig. 4.22.

If the numbers in the tree in Fig. 4.22 all represented a common unit, the tree would be optimal for the data of Table 4.6. However, each number is made up of changes involving different sites. This fact makes the tree suboptimal. In order to show this, consider the tree in Fig. 4.23 in which the changes themselves appear. We use the convention $i\alpha\beta$ to represent a

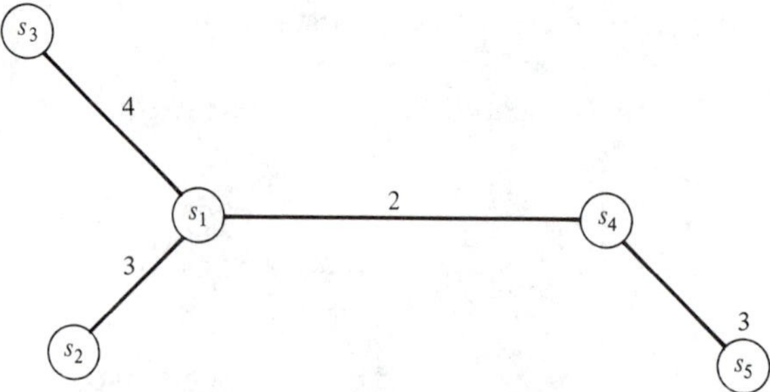

Figure 4.22. The Minimal Spanning Tree.

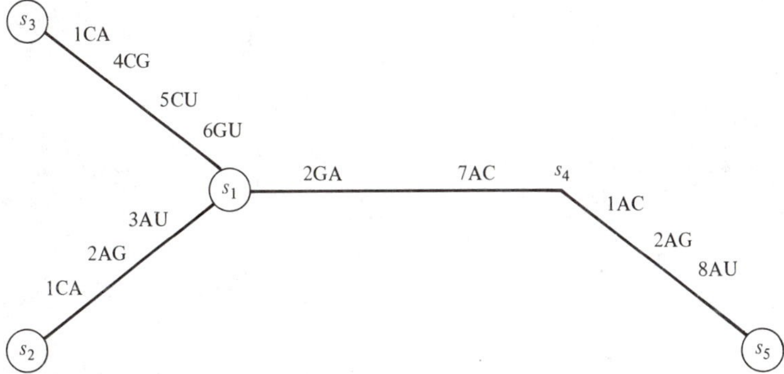

Figure 4.23. The Minimal Spanning Tree with the Changes Indicated.

change at site i from nucleotide α to nucleotide β. As an example, on the s_2s_1 edge one of the three changes is $1CA$, indicating that at site 1, s_2 has a C and s_1 has an A. It is interesting to note that the edges s_1s_4 and s_4s_5 have an identical change, namely, $2GA/2AG$. The two instances of these changes can be combined to produce a new tree, shown in Fig. 4.24, which has one change less. This has created a new vertex in the tree, representing a new sequence s_6. The sequence s_6 is identical to s_4 except that at site 2 it has a G instead of an A and thus $s_6 = AGUGUUCA$. The general reduction process is explained in Fig. 4.25. Suppose two adjacent edges (in this case Pig-Whale, and Dog-Whale) have a change, say $i\alpha\beta$, in common. Then the two instances of this change can be reduced to one by the creation of a new vertex in the tree (here represented by the label "Wig") as illustrated. This process can be repeated on the same pair of edges for further duplicated changes and on other pairs of edges. Indeed the tree in Fig. 4.24 can have its total number of changes reduced from 11 to 10 by applying this *coalescement* process to the duplication of change $1CA$ on adjacent edges s_3s_1 and s_2s_1. This produces the tree shown in Fig. 4.26, which has the new sequence $s_7 = CGUGUUAA$.

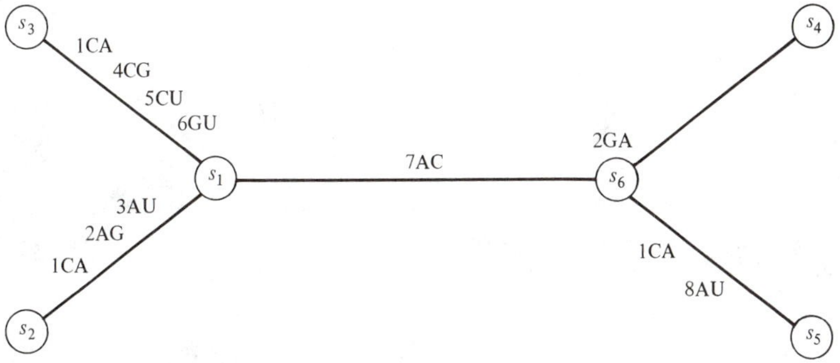

Figure 4.24. The Phylogeny Produced by Coalescement.

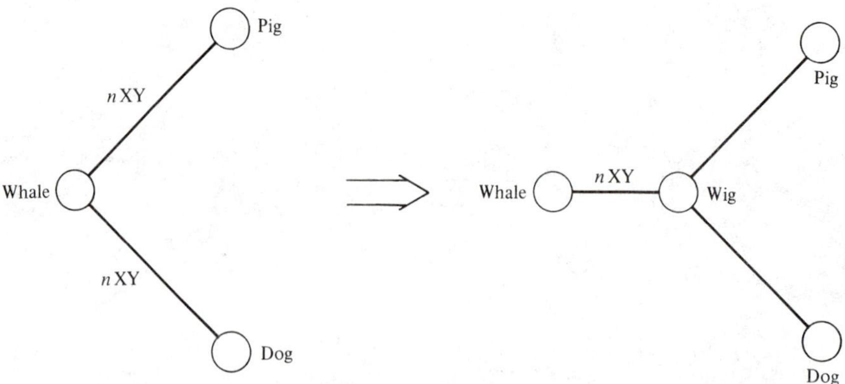

Figure 4.25. The Coalescement Process.

We have seen that the problem is more complicated than the MSTP, as we have found a tree in Fig. 4.26 which has length two changes less than the MSTP tree of Fig. 4.23. This complication is brought about by the fact that the introduction of new sequences (in this case s_6 and s_7) may lead the way to a tree with a smaller number of changes. Vertices representing these new sequences are termed *Steiner points*.

The problem we are considering, that is, constructing a phylogeny of maximum parsimony, has the following formalization:

For a fixed alphabet $\theta = \{A, C, G, U\}$, let d denote the following distance function on θ^N, the set of all sequences of length N on the symbols of θ:

$$d\big((a_1, a_2, \ldots, a_N), (a'_1, a'_2, \ldots, a'_N)\big) = \text{the number of indices } i \text{ such that}$$

$a_i \neq a'_i$, where $(a_1, a_2, \ldots, a_N)$ and $(a'_1, a'_2, \ldots, a'_N)$ are two sequences from θ^N.

Then the *Steiner problem in phylogeny* is, given a set of sequences $S \subseteq \theta^N$, find a Steiner minimal tree for S in θ^N of minimum total length where the length of each edge is defined by d.

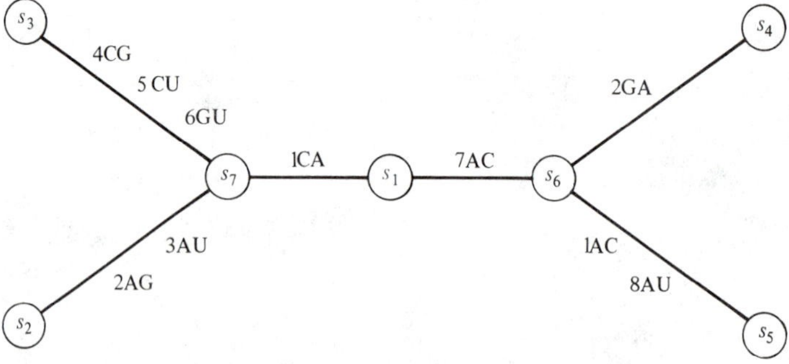

Figure 4.26. The Minimal Phylogeny.

4.5.3. A Solution Procedure for the Steiner Problem in Phylogeny

Unfortunately, the Steiner problem in phylogeny is NP-Complete (see Chapter 2 for an explanation of the term) making the possiblity of an algorithm guaranteeing an optimal solution remote. We now explain a heuristic procedure by using it to solve the numerical example furnished in Table 4.6.

The method begins by creating Table 4.7 from the given set of sequences. A modified version of Prim's method (see Section 3.1.2) is then used to create a spanning tree. Edges are accepted in the order defined by Prim's method, based on Table 4.7. However, after each edge is added a check is made to see whether coalescement is possible. If it is, all possible coalescement is carried out. Choices for coalescement are made on the basis of which will bring about the largest reduction in the total number of changes in the tree constructed so far. When no further coalescement can be carried out, the sequences of the Steiner points created by the coalescement (if any) are deduced. These new sequences are added to Tables 4.6 and 4.7. The cycle of the addition of the next edge via Prim's method and the search for coalescement is then repeated until a tree spanning the given set of species has been created. The heuristic is then terminated. The reader should verify that the sequence of trees given in Figs. 4.23, 4.24, and 4.26 can be produced by this method.

A natural question that arises is whether the final tree in Fig. 4.26 has the minimal possible number of changes. To discover whether or not it is minimal, we analyze the original sequences given in Table 4.6. Consider the subsequences produced by considering sites 1, 2, and 7 in that order, as shown in Table 4.8.

Table 4.8. Subsequences Based on Three Selected Sites

		Sites		
		1	2	7
Species	s_1	A	G	A
	s_2	C	A	A
	s_3	C	G	A
	s_4	A	A	C
	s_5	C	G	C

The final phylogeny produced by the solution procedure just explained, given in Fig. 4.26, then reduces to that shown in Fig. 4.27. It has only one Steiner point as s_7 is identical to s_3 on sites 1, 2, and 7 and has a total of five changes. As each edge has only one change, the introduction of any further Steiner points must lead to a phylogeny with more than five changes.

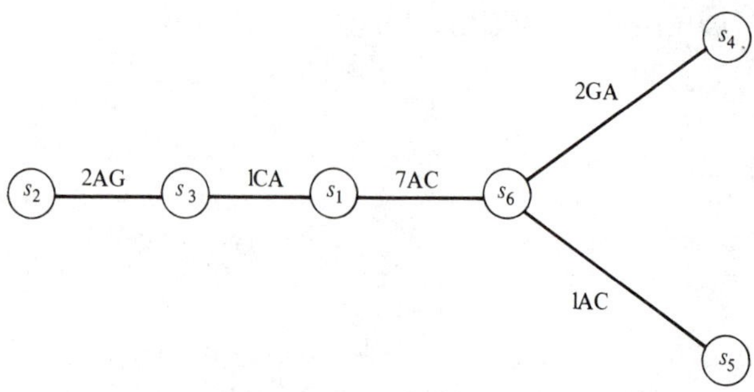

Figure 4.27. The Phylogeny Based on Sites 1, 2, and 7.

Is it possible to create a phylogeny with fewer changes by not allowing any Steiner points: Any such phylogeny can be found as a solution to the appropriate MSTP. To solve the MSTP, the distance table for the data in Table 4.8 is calculated and is shown in Table 4.9. If an MSTP method is applied to these data, the phylogeny shown in Fig. 4.28 can be produced.

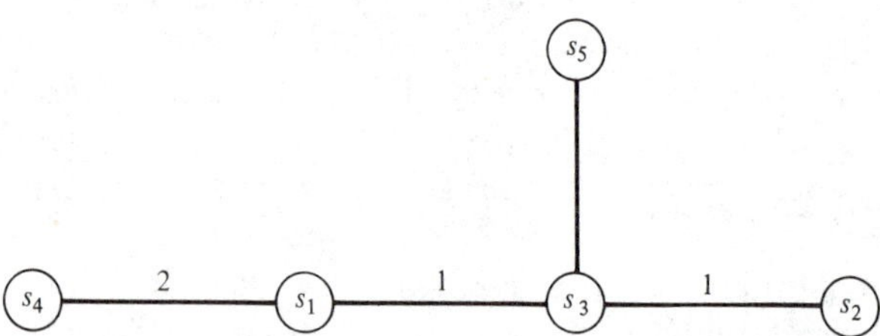

Figure 4.28. The Minimal Spanning Tree for Sites 1, 2, and 7.

It has five changes which is no less than the number in the phylogeny in Table 4.8. Thus we know that at least five changes are required for the three sites 1, 2, and 7 combined. The other five sites 3, 4, 5, 6, and 8, all have two symbols in their columns, as can be seen by studying Table 4.6. (A site which comprises solely a single symbol can be removed as it provides no comparative information.) Thus there must be at least one change somewhere in any phylogeny on these data for each of these sites. Therefore there must be at least five changes in any phylogeny in order to explain these five

Table 4.9. The Distance Table for the Data of Table 4.8

		Species			
		s_2	s_3	s_4	s_5
Species	s_1	2	1	2	2
	s_2		1	2	2
	s_3			3	1
	s_4				2

sites. There must be a further five distinct changes in order to explain the remaining sites 1, 2, and 7. Hence any phylogeny must have at least 10 changes. But the phylogeny in Fig. 4.26 has exactly 10 changes and is therefore a minimal phylogeny. There are other minimal phylogenies and the reader is urged to use the ideas just presented to find them all.

What we have done is to partition the set of sites and solve a Steiner problem in phylogeny for each subset in the partition. This decomposes the original problem into a series of smaller, easier problems. The approach relies upon the Theorem 4.1.

Theorem 4.1. *Let the set of sites* $S=(t_1, t_2, \ldots, t_N)$ *be partitioned into* r *subsets*: $T_1, T_2, \ldots, T_r$. *Let the number of changes of any minimal spanning tree for the subsequences induced by* T_i *be* q_i, $i=1,2,\ldots,r$. *If* L *is the number of changes in any phylogeny of maximum parsimony for* S *then*

$$L \geq \sum_{i=1}^{r} q_i.$$

4.5.4. Exercises

1 Use the solution procedure of Section 4.5.3 to construct a phylogeny for the following data.

		Sites					
		1	2	3	4	5	6
Species	s_1	*A*	*G*	*G*	*C*	*U*	*G*
	s_2	*C*	*G*	*A*	*A*	*U*	*C*
	s_3	*C*	*G*	*A*	*A*	*U*	*U*
	s_4	*A*	*C*	*A*	*A*	*U*	*A*
	s_5	*U*	*C*	*G*	*A*	*A*	*G*

2. Construct the distance table for the data in problem 1 above. Find the minimal spanning tree for these data using Prim's method. Calculate the percentage saving in the total number of changes made by the solution generated in solving problem 1 over the solution just generated.

*3. Show that for any set of sequences a phylogeny where Steiner points are allowed has no less than 50% of the number of changes present in any minimal-spanning-tree solution.

CHAPTER 5

Appendix

5.1. Linear Algebra

5.1.1. Matrices

Definition 5.1. A *matrix* is a rectangular array of real numbers. It is usual to display a matrix $\mathbf{A}$ as

$$\mathbf{A} = \begin{bmatrix} a_{11} & a_{12} & \cdots & a_{1n} \\ a_{21} & a_{22} & \cdots & a_{2n} \\ \vdots & \vdots & & \vdots \\ a_{m1} & a_{m2} & \cdots & a_{mn} \end{bmatrix}.$$

As can be seen, $\mathbf{A}$ has m rows and n columns. To save space, $\mathbf{A}$ is sometimes represented by $(a_{ij})_{m\times n}$.

Definition 5.2. A matrix is termed *square* if $m = n$.

Definition 5.3. The *identity matrix* $\mathbf{I}_n$ or $\mathbf{I}$ (if n is known) is the square matrix $(a_{ij})_{n\times n}$, where

$$a_{ij}\begin{cases} =1 & \text{if } i=j \\ =0 & \text{otherwise.} \end{cases}$$

Definition 5.4. The *zero matrix* $\mathbf{0}_{m\times n}$ or $\mathbf{0}$ (if m and n are known) is the matrix $(a_{ij})_{m\times n}$, where

$$a_{ij} = 0, \qquad i=1,2,\ldots,m,\ j=1,2,\ldots,n.$$

Definition 5.5. The *transpose* $\mathbf{A}^T$ of matrix $\mathbf{A}=(a_{ij})_{m\times n}$ is the matrix $(a^T)_{n\times m}$, where

$$a_{ij}^T = a_{ji}, \qquad i=1,2,\dots,m,\ j=1,2,\dots,n.$$

Definition 5.6. Two matrices $\mathbf{A}=(a_{ij})_{m\times n}$ and $\mathbf{B}=(b_{ij})_{m\times n}$ are said to be *equal* if and only if

$$a_{ij} = b_{ij}, \qquad i=1,2,\dots,m,\ j=1,2,\dots,n.$$

This definition carries the implication that two matrices cannot be equal if they do not have the same dimensions m and n.

5.1.2. Vectors

Definition 5.7. A *vector* is a matrix $\mathbf{V}=(v)_{m\times n}$ in which either of the following holds true:

(a) $m=1$.
(b) $n=1$.

In (a) the vector is called a *row vector* and is denoted by $(v_1, v_2, v_3, \dots, v_n)$. Note that the first subscript (which is always one) has been dropped.

In (b) the vector is called a *column vector* and is denoted by

$$\begin{pmatrix} v_1 \\ v_2 \\ \vdots \\ v_m. \end{pmatrix}.$$

Note that the second subscript (which is always one) has been dropped.

Definition 5.8. A vector which is either a row vector with n rows or a column vector with n columns is called an *n vector*.

5.1.3. Simple Matrix Operations

5.1.3.1. *Scalar Multiplication.* The *scalar product* of a real number (often called a scalar) and a matrix $\mathbf{A}=(a_{ij})_{m\times n}$ is a matrix $\alpha\mathbf{A}=(a'_{ij})_{m\times n}$, where

$$a'_{ij} = \alpha a_{ij}, \qquad i=1,2,\dots,m,\ j=1,2,\dots,n.$$

5.1.3.2. *Addition.* The *sum* of two matrices $\mathbf{A}=(a_{ij})_{m\times n}$ and $\mathbf{B}=(b_{ij})_{m\times n}$ is the matrix $\mathbf{C}=(c_{ij})_{m\times n}$ (denoted by $\mathbf{A}+\mathbf{B}$), where

$$c_{ij} = a_{ij} + b_{ij}, \qquad i=1,2,\dots,m,\ j=1,2,\dots,n.$$

Note that the two matrices cannot be added together unless they have the same dimensions.

5.1.3.3. *Subtraction*. The *difference* between two matrices $\mathbf{A}=(a_{ij})_{m\times n}$ and $\mathbf{B}=(b_{ij})_{m\times n}$ is a matrix $\mathbf{C}=(c_{ij})_{m\times n}$ (denoted by $\mathbf{A}-\mathbf{B}$), where

$$c_{ij}=a_{ij}-b_{ij}, \qquad i=1,2,\ldots,m,\ j=1,2,\ldots,n.$$

As in addition, two matrices cannot be subtracted one from the other unless they have the same dimensions.

5.1.3.4. *Multiplication*. The *vector product* of two matrices $\mathbf{A}=(a_{ij})_{m\times n}$ and $\mathbf{B}=(b_{ij})_{n\times p}$ is a matrix $\mathbf{C}=(c_{ij})_{m\times p}$ (denoted by $\mathbf{AB}$), where

$$c_{ij}=\sum_{k=1}^{n} a_{ik}b_{kj}, \qquad i=1,2,\ldots,m,\ j=1,2,\ldots,p.$$

Note that two matrices cannot be multiplied together in this way unless the first has the same number of columns as the second has rows. Therefore only if $\mathbf{A}$ and $\mathbf{B}$ are both square of the same dimensions is it true that $\mathbf{AB}$ and $\mathbf{BA}$ are both defined. Usually, $\mathbf{AB}\neq\mathbf{BA}$. Here are some examples of matrix multiplication:

$$\begin{pmatrix}4 & 5 & 2\\ 1 & 4 & 6\end{pmatrix}\begin{pmatrix}1 & 4\\ 3 & 3\\ 1 & 7\end{pmatrix}=\begin{pmatrix}4\times1+5\times3+2\times1 & 1\times1+4\times3+6\times1\\ 4\times4+5\times3+2\times7 & 1\times4+4\times3+6\times7\end{pmatrix}$$

$$=\begin{pmatrix}21 & 19\\ 45 & 58\end{pmatrix},$$

$$\begin{pmatrix}1 & 2\end{pmatrix}\begin{pmatrix}8 & 9\\ 10 & 11\end{pmatrix}=\begin{pmatrix}1\times8+2\times10 & 1\times9+2\times11\end{pmatrix}$$

$$=\begin{pmatrix}28 & 31\end{pmatrix},$$

and

$$\begin{pmatrix}12 & 13\end{pmatrix}\begin{pmatrix}2\\ 3\end{pmatrix}=12\times2+13\times3=63 \text{ (a scalar)}.$$

Properties of the Operations

Consider any matrices **A**, **B**, **C**, **D**, and **E** each with m rows and n columns and any real number α. Then

(i) $\mathbf{A}+\mathbf{0}_{m\times n}=\mathbf{0}_{m\times n}+\mathbf{A}=\mathbf{A}$.
(ii) $\mathbf{A}+\mathbf{B}=\mathbf{B}+\mathbf{A}$.
(iii) $\mathbf{A}+(\mathbf{B}+\mathbf{C})=(\mathbf{A}+\mathbf{B})+\mathbf{C}$.
(iv) $\mathbf{A}-(\mathbf{B}-\mathbf{C})=(\mathbf{A}-\mathbf{B})-\mathbf{C}$.
(v) $(\mathbf{A}+\mathbf{B})^T=\mathbf{A}^T+\mathbf{B}^T$.
(vi) $(\mathbf{A}-\mathbf{B})^T=\mathbf{A}^T-\mathbf{B}^T$.
(vii) $\alpha(\mathbf{A}+\mathbf{B})=\alpha\mathbf{A}+\alpha\mathbf{B}$.

Suppose now that the dimensions of the matrices are changed as necessary so that the following multiplications are defined. Then

(viii) $\mathbf{IA} = \mathbf{AI} = \mathbf{A}$.
(ix) $\mathbf{A}(\mathbf{DE}) = (\mathbf{AD})\mathbf{E}$.
(x) $\mathbf{A}(\mathbf{D}+\mathbf{E}) = \mathbf{AD}+\mathbf{AE}$.
(xi) $(\mathbf{D}+\mathbf{E})\mathbf{A} = \mathbf{DA}+\mathbf{EA}$.
(xii) $\alpha(\mathbf{AB}) = (\alpha\mathbf{A})\mathbf{B} - \mathbf{A}(\alpha\mathbf{B})$.

Definition 5.9. A finite set $(\mathbf{V}_1, \mathbf{V}_2, \ldots, \mathbf{V}_q)$ of n vectors is *linearly independent* if and only if for a set of real numbers $\alpha_1, \alpha_2, \ldots, \alpha_q$ such that

$$\sum_{i=1}^{q} \alpha_i \mathbf{V}_i = 0,$$

then $\alpha_1 = \alpha_2 = \cdots = \alpha_q = 0$. (Here $\alpha_i \mathbf{V}_i$ denotes the scalar multiplication of vector $\mathbf{V}_i$ by scalar α_i.)

Definition 5.10. If a finite set of vectors is not linearly independent it is said to be *linearly dependent*.

As an example consider the set $(\mathbf{V}_1, \mathbf{V}_2, \mathbf{V}_3)$ of vectors where $\mathbf{V}_1 = (1,0,0)$, $\mathbf{V}_2 = (0,1,0)$, and $\mathbf{V}_3 = (0,0,1)$. This set is linearly independent because suppose we find three scalars α_1, α_2, and α_3 such that

$$\begin{aligned} \alpha_1\mathbf{V}_1 + \alpha_2\mathbf{V}_2 + \alpha_3\mathbf{V}_3 &= 0. \\ \therefore (\alpha_1,0,0)+(0,\alpha_2,0)+(0,0,\alpha_3) &= 0. \\ \therefore (\alpha_1,\alpha_2,\alpha_3) = \mathbf{0} &= (0,0,0). \\ \therefore \alpha_1 = \alpha_2 = \alpha_3 &= 0. \end{aligned} \tag{5.1}$$

However, if $\mathbf{V}_3$ is $(3,4,0)$ the set is linearly dependent because we can find a set of scalars $(\alpha_1, \alpha_2, \alpha_3)$ which satisfy (5.1) and are not all zero. One such set is

$$\alpha_1 = 3, \qquad \alpha_2 = 4, \qquad \alpha_3 = -1.$$

5.1.4. Determinants

Consider any square matrix $\mathbf{A}$. We can calculate a unique real number for it, called the *determinant* of $\mathbf{A}$, denoted by $|\mathbf{A}|$ or $\det(\mathbf{A})$. We now build up the equations needed to calculate determinants. Consider the matrix derived from $\mathbf{A}$ by deleting the ith row and the jth column of $\mathbf{A}$. The determinant of this new matrix is called the *i–j minor* and is denoted by $\mathbf{M}_{ij}$. We also define the *cofactor*, denoted by C_{ij}, for each i–j pair of rows and columns, as

$$C_{ij} = (-1)^{i+j} M_{ij}. \tag{5.2}$$

We define the determinant of a square matrix $\mathbf{A}=(a_{ij})_{n\times n}$ to be

$$|\mathbf{A}|=\sum_{j=1}^{n} a_{rj}C_{rj}, \tag{5.3}$$

where r denotes any row of $\mathbf{A}$, that is, $1\le r\le n$.

Substituting (5.2) into (5.3):

$$|\mathbf{A}|=\sum_{j=1}^{n} a_{rj}(-1)^{r+j}M_{rj}. \tag{5.4}$$

On looking at (5.4) we see that we need to calculate determinants for the $(n-1)\times(n-1)$ matrices M_{rj}. We can use (5.4) recursively to calculate these, and thus reducing the problem to finding the determinants of $(n-2)\times(n-2)$ matrices. Using 5.4 in this way we eventually reduce the problem to finding the determinants of 2×2 matrices. In the 2×2 case, (5.4) reduces to

$$|(a_{ij})_{2\times 2}|=\left|\begin{pmatrix} a_{11} & a_{12}\\ a_{21} & a_{22}\end{pmatrix}\right|=a_{11}a_{22}-a_{12}a_{21}. \tag{5.5}$$

Let us illustrate the previous ideas by finding the determinant of

$$\mathbf{A}=\begin{pmatrix} 1 & 2 & 3 & 4\\ 5 & 6 & 7 & 8\\ 9 & 10 & 11 & 12\\ 13 & 14 & 15 & 16\end{pmatrix}.$$

Choose $r=1$. Then by (5.4),

$$\begin{aligned}|\mathbf{A}|&=a_{11}(-1)^{1+1}M_{11}+a_{12}(-1)^{1+2}M_{12}\\ &\quad+a_{13}(-1)^{1+3}M_{13}+a_{14}(-1)^{1+4}M_{14}.\\ \therefore |\mathbf{A}|&=M_{11}-2M_{12}+3M_{13}-4M_{14}.\end{aligned} \tag{5.6}$$

We now have to calculate M_{11}, M_{12}, M_{13}, and M_{14}, which are the determinants of the 3×3 matrices:

$$\begin{pmatrix} 6 & 7 & 8\\ 10 & 11 & 12\\ 14 & 15 & 16\end{pmatrix},\quad \begin{pmatrix} 5 & 7 & 8\\ 9 & 11 & 12\\ 13 & 15 & 16\end{pmatrix},\quad \begin{pmatrix} 5 & 6 & 8\\ 9 & 10 & 12\\ 13 & 14 & 16\end{pmatrix},$$

and

$$\begin{pmatrix} 5 & 6 & 7\\ 9 & 10 & 11\\ 13 & 14 & 15\end{pmatrix},$$

respectively.

We can use (5.4) to calculate each of the determinants. For example,

$$\begin{aligned}M_{11}&=6(-1)^{1+1}\left|\begin{pmatrix} 11 & 12\\ 15 & 16\end{pmatrix}\right|+7(-1)^{1+2}\left|\begin{pmatrix} 10 & 12\\ 14 & 16\end{pmatrix}\right|\\ &\quad+8(-1)^{1+3}\left|\begin{pmatrix} 10 & 11\\ 14 & 15\end{pmatrix}\right|.\end{aligned}$$

Each of these 2×2 matrices are now calculated using (5.5).

$$\therefore M_{11}=6(11\times 16-12\times 15)-7(10\times 16-12\times 14)+8(10\times 15-11\times 14)$$
$$=0$$

The others, M_{12}, M_{13}, and M_{14}, are calculated similarly:

$$M_{12}=5(-1)^{1+1}\begin{pmatrix}11 & 12\\ 15 & 16\end{pmatrix}+7(-1)^{1+2}\begin{vmatrix}9 & 12\\ 13 & 16\end{vmatrix}+8(-1)^{1+3}\begin{vmatrix}9 & 11\\ 13 & 15\end{vmatrix}$$
$$=5(11\times 16-12\times 15)-7(9\times 16-12\times 13)+8(9\times 15-11\times 13)$$
$$=0,$$

$$M_{13}=5(-1)^{1+1}\begin{vmatrix}10 & 12\\ 14 & 16\end{vmatrix}+6(-1)^{1+2}\begin{vmatrix}9 & 12\\ 13 & 16\end{vmatrix}+8(-1)^{1+3}\begin{vmatrix}9 & 10\\ 13 & 14\end{vmatrix}$$
$$=5(10\times 16-12\times 14)-6(9\times 16-12\times 13)+8(9\times 14-10\times 13)$$
$$=0,$$

and

$$M_{14}=5(-1)^{1+1}\begin{vmatrix}10 & 11\\ 14 & 15\end{vmatrix}+6(-1)^{1+2}\begin{vmatrix}9 & 11\\ 13 & 15\end{vmatrix}+7(-1)^{1+3}\begin{vmatrix}9 & 10\\ 13 & 14\end{vmatrix}$$
$$=5(10\times 15-11\times 14)-6(9\times 15-11\times 13)+7(9\times 14-10\times 13)$$
$$=0.$$

According to (5.6),

$$|\mathbf{A}|=1(0)-2(0)+3(0)-4(0).$$
$$\therefore |\mathbf{A}|=0.$$

This comes as no surprise as each of the columns is just a multiple of any other. That is they are linearly dependent. The reader should prove this in general and a similar result for rows.

Definition 5.11. If $\mathbf{A}$ is a matrix $(a_{ij})_{m\times n}$, then the *cofactor matrix* of $\mathbf{A}$ is $(\mathbf{C}_{ij})_{m\times n}$, where C_{ij} is the cofactor defined earlier.

Definition 5.12. If $\mathbf{A}$ is a matrix $(a_{ij})_{m\times n}$ then the *adjoint matrix* of $\mathbf{A}$ is the transpose of the cofactor matrix of $\mathbf{A}$, denoted by $\mathbf{A}_{\text{adj}}$.

Properties of Determinants

Let $\mathbf{A}$ be the matrix $(a_{ij})_{m\times n}$.

(i) Consider $\mathbf{B}$ which is obtained from $\mathbf{A}$ by the interchange of two rows (or two columns) of $\mathbf{A}$. Then

$$|\mathbf{A}|=-|\mathbf{B}|.$$

(ii) $|\mathbf{A}|=|\mathbf{A}^T|$.

(iii) If $\mathbf{A}$ has either a row of all zeros or a column of all zeros then $|\mathbf{A}|=0$.

(iv) If a matrix $\mathbf{B}$ is obtained from $\mathbf{A}$ by adding a scalar multiple of a row (column) to another row (column) of $\mathbf{A}$ then $|\mathbf{B}|=|\mathbf{A}|$.

(v) Properties (iii) and (iv) imply that if two rows (or columns) of $\mathbf{A}$ are identical then $|\mathbf{A}|=0$.

(vi) If a matrix **B** is obtained from **A** by multiplying the elements of a row or column of **A** by a scalar then

$$|\mathbf{B}| = \alpha |\mathbf{A}|.$$

(vii) If **A** is square and another matrix has the same dimensions as **A** then

$$|\mathbf{AB}| = |\mathbf{A}|\,|\mathbf{B}|.$$

5.1.5. *Matrix Inverse*

Definition 5.13. If **A** is a square $n \times n$ matrix, then the *inverse* of **A**, denoted by $\mathbf{A}^{-1}$ is the matrix such that

$$\mathbf{A}\mathbf{A}^{-1} = \mathbf{I}_n.$$

Definition 5.14. A matrix **A** is termed *nonsingular* if $|\mathbf{A}| \neq 0$.

Properties of the Matrix Inverse

(i) If **A** is a nonsingular square matrix then $\mathbf{A}^{-1}$ is unique.
(ii) If **A** and **B** are both nonsingular square matrices such that **AB** is defined, then $(\mathbf{AB})^{-1} = \mathbf{B}^{-1}\mathbf{A}^{-1}$.
(iii) If **A** is a nonsingular matrix and **B** and **C** are two matrices such that **AB** and **AC** are defined, then $\mathbf{AB} = \mathbf{AC} \Rightarrow \mathbf{B} = \mathbf{C}$.
(iv) If **A** is a nonsingular matrix then

$$\mathbf{A}^{-1} = \frac{1}{|\mathbf{A}|}(\mathbf{A}_{\text{adj}}).$$

A Numerical Example

In order to illustrate property (iv) let us use it to calculate the inverse of

$$\mathbf{A} = \begin{pmatrix} 1 & 2 & 3 \\ 4 & 5 & 7 \\ 6 & 7 & 9 \end{pmatrix}. \tag{5.7}$$

$$\begin{aligned} |\mathbf{A}| &= 1\times(5\times9-7\times7)-2(4\times9-7\times6)+3(4\times7-5\times6) \\ &= 2. \end{aligned}$$

The cofactor matrix is

$$\begin{pmatrix} -4 & 6 & -2 \\ 3 & -9 & 5 \\ -1 & 5 & -3 \end{pmatrix}.$$

$$\therefore \mathbf{A}_{\text{adj}} = \begin{pmatrix} -4 & 3 & -1 \\ 6 & -9 & 5 \\ -2 & 5 & -3 \end{pmatrix}.$$

$$\therefore \mathbf{A}^{-1} = \begin{pmatrix} -2 & \frac{3}{2} & -\frac{1}{2} \\ 3 & -\frac{9}{2} & \frac{5}{2} \\ -1 & \frac{5}{2} & -\frac{3}{2} \end{pmatrix}.$$

To check:

$$\mathbf{AA}^{-1} = \begin{pmatrix} 1 & 2 & 3 \\ 4 & 5 & 7 \\ 6 & 7 & 9 \end{pmatrix} \begin{pmatrix} -2 & \frac{3}{2} & -\frac{1}{2} \\ 3 & -\frac{9}{2} & \frac{5}{2} \\ -1 & \frac{5}{2} & -\frac{3}{2} \end{pmatrix} = \begin{pmatrix} 1 & 0 & 0 \\ 0 & 1 & 0 \\ 0 & 0 & 1 \end{pmatrix} = \mathbf{I}_3.$$

The inverse of a square nonsingular matrix $\mathbf{A}$ can also be calculated using the *Gauss–Jordon elimination*. If $\mathbf{A}$ is an $n \times n$ matrix we begin by appending $\mathbf{I}_n$ to the right of $\mathbf{A}$ to form $\mathbf{B} = [\mathbf{A}|\mathbf{I}_n]$. $\mathbf{B}$ is now transformed by adding scalar multiples of its rows to other rows of $\mathbf{B}$. This is done in a fashion which produces $\mathbf{I}_n$ in the part of $\mathbf{B}$ occupied by $\mathbf{A}$.

To illustrate this procedure, let us use it to find the inverse of (5.7) (R stands for row):

$$\mathbf{B} = \begin{pmatrix} 1 & 2 & 3 & 1 & 0 & 0 \\ 4 & 5 & 7 & 0 & 1 & 0 \\ 6 & 7 & 9 & 0 & 0 & 1 \end{pmatrix} \quad \begin{matrix} R_1 \\ R_2 \\ R_3 \end{matrix}$$

$$= \begin{pmatrix} 1 & 2 & 3 & 1 & 0 & 0 \\ 0 & -3 & -5 & -4 & 1 & 0 \\ 0 & -5 & -9 & -6 & 0 & 1 \end{pmatrix} \quad \begin{matrix} R_1 \\ R_2 - 4R_1 \\ R_3 - 6R_1 \end{matrix}$$

$$= \begin{pmatrix} 1 & 2 & 3 & 1 & 0 & 0 \\ 0 & 1 & \frac{5}{3} & \frac{4}{3} & -\frac{1}{3} & 0 \\ 0 & -5 & -9 & -6 & 0 & 1 \end{pmatrix} \quad \begin{matrix} R_1 \\ R_2/(-3) \\ R_3 \end{matrix}$$

$$= \begin{pmatrix} 1 & 0 & -\frac{1}{3} & -\frac{5}{3} & \frac{2}{3} & 0 \\ 0 & 1 & \frac{5}{3} & \frac{4}{3} & -\frac{1}{3} & 0 \\ 0 & 0 & \frac{2}{3} & \frac{2}{3} & -\frac{5}{3} & 1 \end{pmatrix} \quad \begin{matrix} R_1 - 2R_2 \\ R_2 \\ R_3 + 5R_2 \end{matrix}$$

$$= \begin{pmatrix} 1 & 0 & -\frac{1}{3} & -\frac{5}{3} & \frac{2}{3} & 0 \\ 0 & 1 & \frac{5}{3} & \frac{4}{3} & -\frac{1}{3} & 0 \\ 0 & 0 & 1 & -1 & \frac{5}{2} & -\frac{3}{2} \end{pmatrix} \quad \begin{matrix} R_1 \\ R_2 \\ R_3/(-\frac{2}{3}) \end{matrix}$$

$$= \begin{pmatrix} 1 & 0 & 0 & -2 & \frac{3}{2} & -\frac{1}{2} \\ 0 & 1 & 0 & 3 & -\frac{9}{2} & \frac{5}{2} \\ 0 & 0 & 1 & -1 & \frac{5}{2} & -\frac{3}{2} \end{pmatrix} \quad \begin{matrix} R_1 + \frac{1}{3}(R_3) \\ R_2 - \frac{5}{3}(R_3^3). \\ R_3 \end{matrix}$$

$\mathbf{A}^{-1}$ is the last three columns of $\mathbf{B}$ which is

$$\mathbf{A}^{-1} = \begin{pmatrix} -2 & \frac{3}{2} & -\frac{1}{2} \\ 3 & -\frac{9}{2} & \frac{5}{2} \\ -1 & \frac{5}{2} & -\frac{3}{2} \end{pmatrix}.$$

5.2. Graph Theory

Over the last 20 years graph theory (GT) has developed into a systematic tool for analyzing practical problems from a wide variety of disciplines. Since its beginning, 250 years ago, remarkably few GT works have used a common terminology. This is not necessarily inefficient as its allows each author to introduce a language tailor-made for his or her purposes. We have chosen a modern terminology here. This section should be used like a dictionary—to be referred to when necessary. It contains a few ideas not referred to earlier. These were included when they arise naturally in the development and may prove useful to the reader if he or she delves more deeply into GT and its applications.

A *graph* G consists of a finite, nonempty set V of *vertices* together with a given set E of unordered pairs of distinct vertices of V. Each element $\{p, q\} \in E$ is called an *edge* and is said to *join* vertices p and q. Synonyms for "vertex" are *point*, *node*, and *junction*. Synonyms for "edge" are *line*, *arc*, *branch*, and *link*. If edge $e = \{p, q\} \in E$, p and q are both said to be *incident* with e and *adjacent* to each other. An edge $\{p, q\}$ is denoted by pq when ambiguity does not arise.

One of the features that makes GT more fun is our ability to represent any graph by a picture. This is always theoretically possible as V is assumed finite. Each vertex in V can be represented by a geometric point in the plane and each edge in E by a geometric line joining the geometric points with which it is incident. As an example, the graph $G = (V, E)$, where $V =$

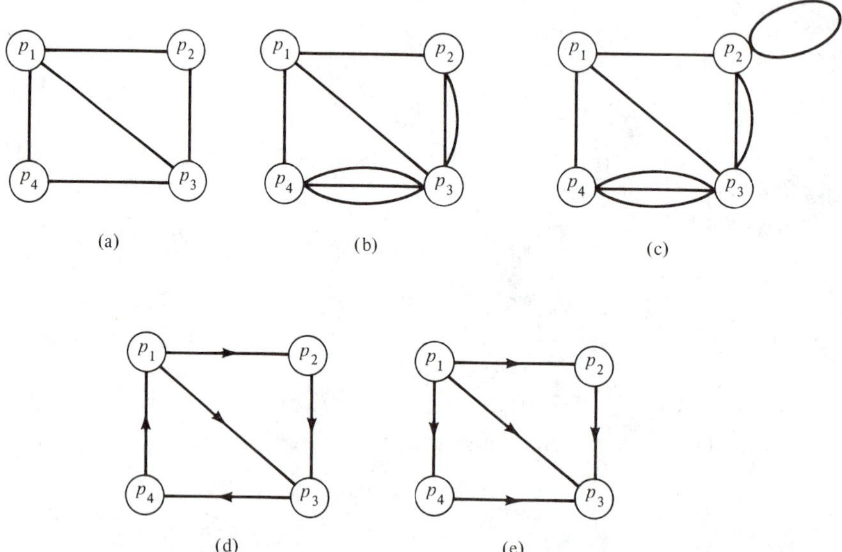

Figure 5.1. (a) Graph; (b) Multigraph; (c) Pseudograph; (d) Digraph; (e) Network.

$\{p_1, p_2, p_3, p_4\}$ and $E = \{\{p_1, p_2\}, \{p_2, p_3\}, \{p_3, p_4\}, \{p_4, p_1\}, \{p_1, p_3\}\}$ is depicted in Fig. 5.1(a).

There are a number of useful structures that bear a close resemblance to the structure of a graph. Recall that no more than one edge can join any two vertices of a graph. Structures in which this constraint is relaxed are called *multigraphs*. Thus a *multigraph* is an ordered pair of sets (V, E), where V is finite and nonempty and E is a class of unordered pairs of distinct vertices of V where repetitions in the class are allowed. An example of a multigraph is given in Fig. 5.1(b). Recall that each edge in a graph and a multigraph joints two *distinct* vertices. Structures in which this constraint is relaxed and multiple edges are allowed are called *pseudographs*. Thus a *pseudograph* is an ordered pair of sets (V, E), where V is finite and nonempty and E is a class of unordered pairs of vertices of V where repetitions in the class are allowed. An example of a pseudograph is given in Fig. 5.1(c). Recall that each edge in a graph is represented by an *unordered* pair of vertices of V. Sometimes it is useful to give the edges an orientation or direction. When this is carried out, the graph is termed *directed* and is called a *digraph*. Thus a *digraph* is an ordered pair of sets (V, A), where V is finite and nonempty and A is a set of ordered pairs of distinct vertices of V. An example of a digraph is given in Fig. 5.1(d). Much graph theoretic terminology can and will be used analogously for digraphs in this chapter. There is a special class of digraphs called *networks* which are of extreme importance. They can be introduced as follows. If $a = (p_i, p_j)$ is an arc in a digraph $D = (V, A)$, then a is said to be *directed away* from p_i and *directed toward* p_j. The vertices of a network are usually called nodes. Any node which has at least one arc directed away from it is said to be a *source*. Any node which has at least one arc directed toward it is said to be a *sink*. Of course nodes can have arcs directed both toward and away from them making them both sources and sinks. A source (sink) which is not a sink (source) is termed a *proper source* (*sink*). A digraph through the arcs of which a commodity is assumed to flow and which usually has at least one proper source and at least one proper sink is termed a *network*. An example of a network is given in Fig. 5.1(e).

Returning to graphs, it is sometimes appropriate to examine just a part of a graph. This can be done in a number of different ways. Given a graph $G = (V, E)$ we could consider: (1) just some of its vertices and all the existing edges in E joining them; (2) all of its vertices and just some of the existing edges joining them; or (3) some of its vertices and some of the edges joining them provided we include all vertices incident with any edges considered.

As an example:

(1) Let U be a nonempty subset of V. The graph whose vertex set is U and whose edge set comprises exactly the edges of E which join vertices in U, is termed a *subgraph* of G.
(2) Let F be a subset of E. (V, F), the graph with the same vertex set as G, but only the edges in F, is termed a *partial* or *spanning graph* of G.

(3) A graph formed by the combination of (1) and (2) where care is taken to include the end points (incident vertices) of each edge selected is termed a *partial subgraph* of G.

Figures 5.2(a)–5.2(c) provide examples of a subgraph, partial graph, and partial subgraph, respectively, of the graph in Fig. 5.2(a).

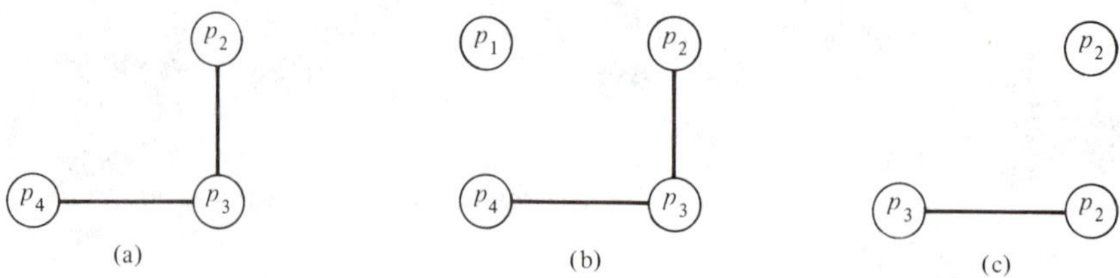

Figure 5.2. Parts of the Graph in Fig. 5.1(a): (a) Subgraph; (b) Partial Graph; (c) Partial Subgraph.

One of the most important graph theoretic concepts is that of connectivity. A *walk* in a graph G is a sequence of vertices and edges of G of the form:

$$\langle p_1, \{p_1, p_2\}, p_2, \{p_2, p_3\}, p_3, \ldots, p_{n-1}, \{p_{n-1}, p_n\}, p_n \rangle.$$

Note that the sequence begins and ends with a vertex and each edge is incident with the vertices immediately preceding and succeeding it. It is said to *join* p_1 and p_n. When there is no ambiguity, it is denoted by $\langle p_1, p_2, \ldots, p_n \rangle$. A walk is termed *closed* if $p_1 = p_n$ and *open* if $p_1 \neq p_n$. A walk is termed a *trail* if all of its edges are distinct and a *path* if all its vertices (and necessarily its edges) are distinct. A closed walk with at least three vertices and all of its vertices distinct is called a *cycle*. A graph G is *connected* if every pair of vertices are joined by at least one path and k *connected* if at least k vertices must be removed to make G no longer connected. In Fig. 5.1(a), $\langle p_1, p_2, p_3, p_1, p_2 \rangle$ is an open walk; $\langle p_1, p_2, p_3, p_4, p_1, p_3 \rangle$ is a trail; $\langle p_1, p_2, p_3 \rangle$ is a path; and $\langle p_1, p_2, p_3, p_4, p_1 \rangle$ is a cycle. A connected partial subgraph of a graph is termed a *spanning subgraph*.

A diagraph is termed *connected* if the underlying graph derived from it by ignoring the orientation of its arcs and replacing multiple edges by a single edge is connected. A digraph is termed *strongly connected* if there is a directed sequence of arcs from each of its vertices to each other vertex. A strongly connected subdigraph d of a digraph D is said to be *minimal* if d has no proper subdigraph of two or more vertices which is strongly connected.

A graph without cycles is termed *acyclic*. This leads us to one of the most important classes of graphs. A connected acyclic graph G is termed a *tree*. There are many equivalent definitions of a tree including:

(1) Every two vertices of G are joined by a unique path.
(2) G is connected and its number of vertices exceeds its number of edges by one.

(3) G is acyclic and its number of vertices exceeds its number of edges by one.
(4) G is acyclic and the addition of a new edge creates exactly one cycle.

The mathematician Leonhard Euler founded the subject of graph theory when, in 1736, he settled what is known as the "Königsberg bridge problem." Until that time, on Sunday afternoons the good citizens of Königsberg (a town in what was then known as Prussia) would amuse themselves in their park. Through the park flowed the lovely river Pregel in which lay two islands, connected by a system of bridges, as shown in Fig. 5.3(a). The challenge was to begin on any of the four pieces of land, walk across each bridge exactly once, and return to the starting point. Euler answered the question as to whether the feat could be achieved for this, or any other system, by modeling the situation in graph theoretic terms. Each piece of land is represented by the vertex of a graph and each bridge by an edge joining the vertices representing the land pieces it connects. This graph G is shown in Fig. 5.3(b). The problem now reduces to settling whether or

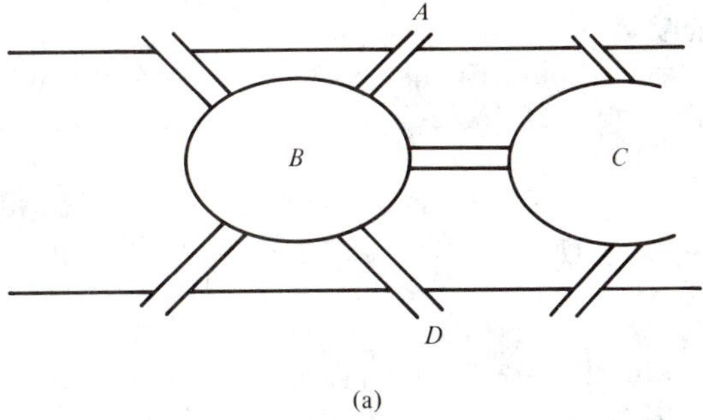

(a)

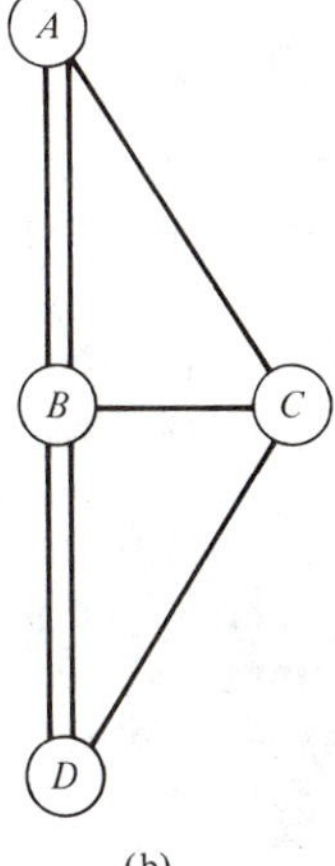

(b)

Figure 5.3. (a) The Königsberg Bridge System; (b) Its GT Model.

not the graph G has the following property: There exists in G a closed walk containing all the edges exactly once each.

A graph with this property is termed *Eulerian* after the great man. Euler proved that the following are equivalent to a graph G being Eulerian (where the *degree* of a vertex v is the number of edges with which it is incident denoted by $d(v)$):

(1) Every vertex of G has even degree.
(2) The set of edges of G can be partitioned into cycles.

It is obvious that the graph in Fig. 5.3(b) does not possess property (1) and thus the desired walk cannot be found.

In 1859 the Astronomer-Royal of Ireland, Sir William Hamilton, invented the following two-player game. He labeled each of the vertices of a certain graph with the name of a city (Amsterdam, Berlin, Edinburgh, etc). The objective of one player was to find a "tour round the world" by finding a cycle of edges that passes through all the cities exactly once each and returns to the starting city. The other player attempted to make this impossible by being allowed periodically to remove certain edges. It turns out that Hamilton's graph does possess a cycle of the required type. Such a graph is termed *Hamiltonian*. Although no elegant characterization of Hamiltonian graphs similar to that given for Eulerian graphs exists, there are available a number of necessary and sufficient conditions (see Harary, 1969, p. 67).

A *bipartite* graph $G=(V,E)$ is a graph for which V can be partitioned into two nonempty, disjoint subsets V_1 and V_2 such that every edge in E joins a vertex in V_1 and a vertex in V_2.

If an edge $e=\{p_1,p_2\}$ exists in a graph, e is said to *cover* p_1 and p_2 and vice versa. It is natural in some circumstances to ask the following question. Given a connected graph, which is the smallest set S of its vertices with the property that every vertex of G is either in S or adjacent to a point in S? Such a set is said to be a *dominating set*. There is another question that could be asked: Which is the largest set S of points with the property that no two vertices in S are adjacent? Such a set is said to be a *maximal independent set*. A *matching* is a subset M of E with the property that no two edges of M have a vertex in common. If M covers V, that is, all the vertices of V are paired off by M, M is said to be *perfect*.

A graph is said to be *embedded* in a plane when it is drawn in that plane in such a way that no two edges intersect. A graph is termed *planar* if it can be embedded in the plane. A graph is termed *maximally planar* if no further edges can be added to it without destroying its planarity. A number of characterizations of planar graphs are given in Harary (1969, Ch. 11).

Until 1976 the most famous unsolved problem in mathematics was the *Four Colour Conjecture*. It states:

> Any map on a plane can be coloured with only four colours so that no two adjacent regions have the same colour.

It is a GT problem because every map can be represented by a graph in which regions are represented by vertices and two vertices are joined by an edge whenever the corresponding regions are adjacent. It is always possible to find a planar graph to represent the map in this way. Thus the problem reduces to proving that the vertices of a planar graph can be colored with four colors in such a way that every pair of adjacent vertices have different colors. This conjecture was proved in the affirmative in 1976 by Haken and Appel. Even so there are a number of interesting questions left concerning the coloring of the vertices of graphs. The minimum number of colors required to color the vertices of a graph (with adjacent vertices having different colors) is termed the *chromatic number* of the graph.

It is often useful to be able to summarize the structure of a graph in matrix form. There are several matrices which completely specify a graph:

(1) The *adjacency matrix* $\mathbf{A}=(a_{ij})$ of a graph with p vertices is a $p\times p$ symmetric matrix in which

$$a_{ij}\begin{cases}=1 & \text{if vertices } p_i \text{ and } p_j \text{ are adjacent}\\ =0 & \text{otherwise.}\end{cases}$$

The adjacency matrix for the graph in Fig. 5.1(a) is

$$\begin{pmatrix}0 & 1 & 1 & 1\\ 1 & 0 & 1 & 0\\ 1 & 1 & 0 & 1\\ 1 & 0 & 1 & 0\end{pmatrix}.$$

Obviously each row sum is the degree of the corresponding vertex. Further, the $i-j$ entry in $\mathbf{A}^n$ is the number of walks with n edges from p_i to p_j. The adjacency matrix of a digraph $D=(V,A)$ can be similarly defined with

$$a_{ij}\begin{cases}=1 & \text{if arc } (p_i,p_j)\in A\\ =0 & \text{otherwise.}\end{cases}$$

Thus this latter matrix is not in general symmetric.

(2) The *incidence matrix* $\mathbf{B}=[b_{ij}]$, in which it is assumed that edges as well as vertices of the graph G are labeled, is a $p\times q$ matrix in which

$$b_{ij}\begin{cases}=1 & \text{if vertex } p_i \text{ and edge } e_j \text{ are incident}\\ =0 & \text{otherwise,}\end{cases}$$

where

$$q=\text{the number of edges of } G.$$

Finally, a *weighted graph* is an ordered pair (G,w) where G is a graph (V,E) and w is a function $w\colon E\to\mathbb{R}$ which maps the edges of G into the set of real numbers. That is, in a weighted graph each edge $\{p_i,p_j\}$ has a weight $w(\{p_i,p_j\})$, usually denoted by w_{ij}. The *weight of a graph* is defined to be the sum of the weights of its edges.

Further Reading

General

There are numerous books in operations research, management science, engineering, mathematics, and computer science that contain substantial material on combinatorial optimization. Also there are at least two graduate-level text books dealing solely with combinatorial optimization:

Lawler, E. L. Combinatorial Optimization: *Networks and Matroids*. New York: Holt, Rienhart, and Winston, 1976.

Papadimitriou, C., and Steiglitz, K. *Combinatorial Optimization: Algorithms and Complexity*. Englewood Cliffs, N.J.: Prentice-Hall, 1982.

Chapter 0

The material on the shortest Hamiltonian path problem is based on:

Robinson, D. F. *Proceedings of the First Australian Conference on Combinatorics*, Newcastle, Australia, 1972.

The following text on general optimization contains more details on the basic concepts of optimization:

Foulds, L. R. *Optimization Techniques: An Introduction*. New York: Springer-Verlag, 1981.

Chapter 1

The classic text on linear programming was written by the inventor of the simplex method:

Dantzig, G. B. *Linear Programming and Extensions*. Princeton, N.J.: Princeton Univ. Press, 1963.

Among the many other excellent books on linear programming we recommend three. The first is at an elementary level, the second intermediate, and the third advanced.

Daellenbach, H. G., and Bell, E. J. *Users Guide to Linear Programming*. Englewood Cliffs, N.J.: Prentice-Hall, 1970.
Bazaraa, M. S., and Jarvis, J. J. *Linear Programming and Network Flows*. New York, Wiley, 1971.
Gass, S. I. *Linear Programming*. 4th ed. New York: McGraw-Hill, 1975.

Chapter 2

Section 2.1

The following is a very readable text on integer programming:

Taha, H. A. *Integer Programming*. New York: Macmillan, 1978.

The following two books cover integer programming at an intermediate and advanced level, respectively:

Greenberg, N. *Integer Programming*. New York: Academic Press, 1971.
Garfinkel, R. S., and Nemhauser, G. L. *Integer Programming*. New York: Wiley, 1972.

A good introduction to branch-and-bound enumeration with applications to a number of problems, including the traveling-salesman problem is:

Lawler, E. L., and Wood, D. E. "Branch and Bound Methods: A Survey." *Oper. Res*. **14** (1966): 699–719.

The branch-and-bound method for the general integer-programming problem presented here is based on:

Dakin, R. J. "A Tree-Search Algorithm for Mixed Integer Programming Problems." *Comp. J.* **8**, no. 3 (1965): 250–255.

The zero–one method is based on:

Balas, E. "An Additive Algorithm for Solving Linear Programs with Zero–One Variables." *Oper. Res.* **13** (1965): 517–546.

The cutting-plane method is based on:

Gomory, R. E. "Outline of an Algorithm for Integer Solutions to Linear Programs." *Bull. Am. Math. Soc.* **14** (1958): 275–278.

Section 2.2

The first book on dynamic programming was written by its inventor:

Bellman, R. E. *Dynamic Programming.* Princeton, N.J.: Princeton Univ. Press, 1957.

One of the most easily read books on dynamic programming is:

Dreyfus, S. E., and Law, A. M. *The Art and Theory of Dynamic Programming.* New York, Academic Press, 1977.

Other books on dynamic programming, at a more advanced level, include:

Bellman, R. E., and Dreyfus, S. E. *Applied Dynamic Programming.* Princeton, N.J.: Princeton Univ. Press, 1962.
Hadley, G. *Nonlinear and Dynamic Programming.* Reading, Mass.: Addison-Wesley, 1964.
Nemhauser, G. L. *Introduction to Dynamic Programming.* New York: Wiley, 1966.
White, D. J. *Dynamic Programming.* Edinburgh: Oliver and Boyd, 1969.

Section 2.3

A thorough introduction to computer algorithms is given in:

Aho, A. V., Hopcroft, J. E., and Ullman, J. D. *The Design and Analysis of Computer Algorithms*, Reading, Mass.: Addison-Wesley, 1974.

One of the first articles to explicitly mention polynomially time-bounded algorithms was:

Edmonds, J. "Paths, Trees, and Flowers," *Can. J. Math.* **17** (1965): 449–467.

The theory of NP-Completeness began with the following paper:

Cook, S. A. "The Complexity of Theorem Proving Procedures." *Proc. 3rd ACM Symp. Theory Comput.* **ACM** (1971): 151–158.

Cook's work was interpreted and many of its consequences were made known in the following two papers:

Karp, R. M. "Reducibility among Combinatorial Problems." In R. E. Miller and J. W. Thatcher (Eds.), *Complexity of Computer Computations*. New York: Plenum, 1972, pp. 85–103.

Karp, R. M. "On the Complexity of Combinatorial Problems." *Networks* **5** (1975): 45–68.

There is also an encyclopedia of NP-Complete problems:

Garey, M. R., and Johnson, D. S. *Computers and Intractability: A Guide to the Theory of NP-Completeness*. San Francisco: Freeman, 1979.

The general approach of heuristic problem solving is discussed in:

Foulds, L. R. "The Heuristic Problem Solving Approach." *J. Oper. Res. Soc.* **34** (1983): 927–934.

Mueller-Merbach, H. "Heuristics and Their Design: A Survey," *European J. of Opl. Res.* **8** (1981): 1–23.

Catalogs of heuristic methods appear in the following articles:

Garey, M. R., and Johnson, D. S. "Approximation Algorithms for Combinatorial Problems: An Annotated Bibliography." In J. F. Traub (Ed.), *Algorithms and Complexity: New Directions and Recent Results*. New York: Academic Press, 1976, pp. 41–52.

Chapter 3

Section 3.1

The two minimal-spanning-tree algorithms are based on:

Kruskal, J. B. "On the Shortest Spanning Subtree of a Graph and the Traveling Salesman Problem." *Proc. Am. Math. Soc.* **7** (1956): 48–50.

Prim, R. G. "Shortest Connection Networks and Some Generalizations." *Bell Systems Technical Journal* **36** (1957): 1389–1401.

Section 3.2

The two shortest-path algorithms are based on:

Dijkstra, E. W. "A Note on Two Problems in Connection with Graphs." *Numerische Mathematik* **1** (1959): 269–271.

Floyd, R. W. "Algorithm 97: Shortest Path." *Communications of the Association for Computation Machinery* **5** (1962): 345.

This last article is cryptic to say the least. A less obscure reference for this algorithm has been written by a coinventor:

Warshall, S. "A Theorem on Boolean Matrices." *Journal of the Association of Computational Machinery* **9** (1962): 11–12.

Section 3.3

The classic text on the theory of network flows is:

Ford, L. R., and Fulkerson, D. R. *Flows in Networks*. Princeton, N.J.: Princeton Univ. Press, 1962.

There are a number of other texts on networks including:

Hu, T. C. *Integer Programming and Network Flows*. Reading, Mass.: Addison-Wesley, 1969.

Minieka, E. *Optimization Algorithms for Networks and Graphs*. New York: Marcel Dekker, 1978.

Smith, D. K. *Network Optimization Practice: A Computational Guide*. Chichester, England: Ellis Horwood, 1982.

This latter text contains a number of BASIC and PASCAL computer programs for various network algorithms.

The labeling method is based on:

Ford, L. R., and Fulkerson, D. R. "Maximal Flow Through a Network." *Can. J. Math*. **8** (1956): 399–404.

Section 3.4

The algorithm for the minimal-cost-flow problem is based on:

Busacker, R. G., and Gowan, P. J. *A Procedure for Determining a Family of Minimal-Cost Network Flow Patterns*." ORO Technical Report No. 15, Operations Research Office, John Hopkins University, Baltimore, Maryland, 1961.

There are more efficient algorithms for the minimal-cost-flow problem. Two are detailed in:

Fulkerson, D. R. "An Out-of Kilter Method for Minimal Cost Flow Problems." *Journal of the Society for Industrial and Applied Mathematics* **9** (1961): 18–27.

Klein, M. "A Primal Method for Minimal Cost Flows." *Manage. Sci*. **14** (1967): 205–220.

Section 3.5

The activity-node approach for activity network analysis is detailed in:

Robinson, D. F., and Foulds, L. R. *Digraphs: Theory and Techniques*. New York: Gordon and Breach, 1980.

The activity-arc approach can be found in:

Lockyer, K. G. *Introduction to Critical Path Analysis*. London, England: Pitman, 1978.

Chapter 4

Section 4.1

The facilities layout problem and its techniques have been surveyed in:

Foulds, L. R. "Techniques for Facilities Layout: Deciding which Pairs of Activities Should be Adjacent." *Manage. Sci.* **29** (1983): 1414–1426.

Section 4.2

There are numerous articles on the traveling-salesman problem. An excellent source is the book by Papadimitriou and Steiglitz referenced under the heading "General." For heuristics for the problem refer to:

Rosencrantz, D. J., Stearns, R. E., and Lewis, P. M. "An Analysis of Several Heuristics for the Traveling Salesman Problem." *Journal of the Society of Industrial and Applied Mathematics on Computing* **6** (1977): 563–581.

Section 4.3

Recent work on vehicle scheduling has been surveyed in:

Watson-Gandy, C., and Foulds, L. R. "The Vehicle Scheduling Problem: A Survey." *New Zeal. Oper. Res.* **9** (1981): 73–92.

Section 4.4

This section is based on:

Foulds, L. R., and Robinson, D. F. "The Car Pool Problem." *New Zeal. Oper. Res.* **5** (1977): 101–117.

Section 4.5

A book on the construction of phylogenies is:

Penny, E. D., Hendy, M. D., and Foulds, L. R. *Evolutionary Tree Construction*. New York: Springer-Verlag, to appear.

A paper containing many useful references to this topic is:

Penny, E. D., Foulds, L. R., and Hendy, M. D. "Testing the Theory of Evolution by Comparing Phylogenetic Trees Constructed from Five Different Protein Sequences." *Nature* **297** (1982): 197–200.

A proof that the Steiner problem in phylogeny is NP-Complete appears in:

Foulds, L. R., and Graham, R. L. "The Steiner Problem in Phylogeny is NP-Complete." *Adv. Appl. Math.* **3** (1982): 43–49.

Section 5.1

Many of the books referenced earlier on linear programming contain extensive treatments of linear algebra.

Section 5.2

There are many excellent texts on graph theory, including:

Bondy, J. A., and Murty, U. S. R. *Graph Theory with Applications*. New York: Macmillan, 1976.
Deo, N. *Graph Theory with Applications*, Englewood Cliffs, N.J.: Prentice-Hall, 1974.
Harary, F. *Graph Theory*, Reading, Mass.: Addison-Wesley, 1969.

The textbook referenced for Section 3.5 is concerned solely with digraphs. There is also a paper surveying the applications of graph theory in operations research:

Foulds, L. R. "Graph Theory: A Survey of its Use in Operations Research." *New Zeal. Oper. Res.* **10** (1982): 35–65.

The proof of the Four Colour Theorem can be found in:

Haken, W., and Appel, K. I. "Every Planar Map is Four Colorable," *Bull. Am. Math. Soc.* **82** (1976): 711–712.

Index

Undergraduate Texts in Mathematics

continued from ii

Malitz: Introduction to Mathematical Logic: Set Theory - Computable Functions - Model Theory.
1979. xii, 198 pages. 2 illus.

Martin: The Foundations of Geometry and the Non-Euclidean Plane.
1975. xvi, 509 pages. 263 illus.

Martin: Transformation Geometry: An Introduction to Symmetry.
1982. xii, 237 pages. 209 illus.

Millman/Parker: Geometry: A Metric Approach with Models.
1981. viii, 355 pages. 259 illus.

Owen: A First Course in the Mathematical Foundations of Thermodynamics
1984. xvii, 178 pages. 52 illus.

Prenowitz/Jantosciak: Join Geometrics: A Theory of Convex Set and Linear Geometry.
1979. xxii, 534 pages. 404 illus.

Priestly: Calculus: An Historical Approach.
1979, xvii, 448 pages. 335 illus.

Protter/Morrey: A First Course in Real Analysis.
1977. xii, 507 pages. 135 illus.

Ross: Elementary Analysis: The Theory of Calculus.
1980. viii, 264 pages. 34 illus.

Sigler: Algebra.
1976. xii, 419 pages. 27 illus.

Simmonds: A Brief on Tensor Analysis.
1982. xi, 92 pages. 28 illus.

Singer/Thorpe: Lecture Notes on Elementary Topology and Geometry.
1976. viii, 232 pages. 109 illus.

Smith: Linear Algebra.
1978. vii, 280 pages. 21 illus.

Smith: Primer of Modern Analysis
1983. xiii, 442 pages. 45 illus.

Thorpe: Elementary Topics in Differential Geometry.
1979. xvii, 253 pages. 126 illus.

Troutman: Variational Calculus with Elementary Convexity.
1983. xiv, 364 pages. 73 illus.
Whyburn/Duda: Dynamic Topology.
1979. xiv, 338 pages. 20 illus.

Wilson: Much Ado About Calculus: A Modern Treatment with Applications Prepared for Use with the Computer.
1979. xvii, 788 pages. 145 illus.